Why Don't S[piders Stick]
to Their [Webs]

"This year, the scientifically curi[ous...]
to many hundreds of conundrums.[...]

Financial Times

"Matthews has done a heroic job in revealing the science behind a broad and entertaining range of questions. Ideal fodder for the curious mind."

Roger Highfield, author of *Can Reindeer Fly? The Science of Christmas* and *The Science of Harry Potter: How Magic Really Works*

"[Matthews] takes things that seem hopelessly complicated and explains how simple they are."

Ian Stewart, author of *Professor Stewart's Cabinet of Mathematical Curiosities*

"Robert Matthews is one of the best science writers around."

Duncan Watts, Principal Research Scientist at Yahoo!

"A good mix of questions and answers, entertainment and education, it will make an excellent 'thinkers' Christmas book."

Publishing News

The Sunday Telegraph

Why Don't Spiders Stick to Their Webs?

and 317 other everyday mysteries of science

Robert Matthews

ONEWORLD

A Oneworld Book

First published as *Q & A* by Oneworld Publications in 2005
First published as *Why Don't Spiders Stick to Their Webs?* in 2007
This edition published 2011

ISBN 978-1-85168-900-2

Typeset by Jayvee, Trivandrum, India
Cover design by Dan Mogford
Interior illustrations copyright © Jolyon Troscianko 2007
Printed and bound in Denmark by Nørhaven

Oneworld Publications
185 Banbury Road
Oxford, OX2 7AR
England

Learn more about Oneworld. Join our mailing list to
find out about our latest titles and special offers at:

www.oneworld-publications.com

To Auriol, Ben and Theo

Contents

Preface

People have the wrong idea about science – not least scientists themselves. For the last few hundred years, a story has been doing the rounds to the effect that scientific discoveries are made by first formulating an hypothesis, then performing an experiment, and finally drawing a conclusion. The reality could hardly be more different. Some of the greatest discoveries – radioactivity, genetics, quantum theory – began with experiments whose outcome defied expectation. Others began with grand conclusions about how the universe is put together, with no clue about how to check it experimentally.

But many, perhaps even most, of the great discoveries in science began with a *question*. When Newton saw an apple fall to the ground in the garden of his mother's house (a story which Newton insisted was true), he asked himself how this could happen, and was duly rewarded by the discovery of the universal law of gravitation. When Einstein asked himself as a teenager what it would be like to ride upon a light-beam, his answer led directly to his Special Theory of Relativity, $E = mc^2$ and all that. The American physicist Richard Feynman claimed that his bafflement over the rate of wobble of a dinner plate spinning through the air in a cafeteria ultimately led him to Nobel Prize-winning discoveries about sub-atomic particles.

Preface

Great minds, great discoveries – but, on the face of it, rather trivial questions. The thing is, Nature herself doesn't know the meaning of the word trivial. From the birth of a spiral galaxy to water gurgling down a plughole – all are manifestations of the primordial laws of physics. And time and again the history of science has shown that the key to understanding the universe often lies in asking a great question.

Over several years, I had the privilege of being asked to investigate a host of wonderfully varied questions about life, the universe and everything by readers of the *Sunday Telegraph*. From the origin of blue moons to the origin of the universe, the causes of tides to the fate of odd socks, they came my way in droves each week, my only regret being an inability to take on them all. Instead, I had to be selective, choosing ones whose answers are little-known, counter-intuitive, or have rather deeper implications than one might expect.

This book represents a selection of the many hundreds of questions I received over the years, whose answers I hope you will find especially entertaining and informative. Some deal with fundamental issues about the nature of reality and the limits to knowledge. Others deal with rather more run-of-the-mill matters – like how best to remove ice from your car windscreen, and whether milk should go in before or after the tea.

Whether your taste is for the cosmic or the quotidian, what follows should convince you that the greatest myth of all is that science is merely what men in lab coats do for a living.

Robert Matthews

Acknowledgements

My first debt of gratitude is to Dominic Lawson, erstwhile Editor of the *Sunday Telegraph*, for suggesting the idea of a weekly section of science-related questions and answers. Were it not for his irritating refusal to accept hand-waving arguments and jargon, compiling it would have been a breeze. I was initially dubious about whether there would be enough suitable questions each week to sustain a weekly column, but I reckoned without the insatiable curiosity of *Sunday Telegraph* readers, for which I am especially grateful.

I would also like to thank Victoria Roddam, formerly of Oneworld Publications, and Morven Knowles at Telegraph Books for their enthusiasm about bringing the answers contained herein to a wider audience.

Finally, I am indebted to Fiona Bacon for selecting the questions for inclusion in this book, spotting blunders, categorizing the answers and generally doing many tasks that would have driven me nuts.

Chapter 1

Mysteries of everyday life

? Why is the place you want on a map so often at the edge?

Whenever I encounter one of these little frustrations of everyday life, I apply the maxim of Ian Fleming's character Goldfinger: once is happenstance, twice is a coincidence but three times is enemy action. That is, if the little irritation in question crops up often enough to make you suspect there is some malign force at work, you are probably right. The frustration of finding the place one is looking for in awkward parts of the map is a case in point: it happens so often that it looks very much like a manifestation of Murphy's Law, according to which "If something can go wrong, it will". This suspicion is confirmed by some simple school geometry. Picture a square map, with the "awkward bit" being the strip-like region around its perimeter. Surprisingly, even if the width of this strip-like region is just one-tenth that of the whole map, it mops up a prodigious thirty-six per cent of the total area. Thus, every time you look for some location of such a map, there is a better than 1 in 3 chance it will turn up in that bit around the edge. What fools us is the fact that although it looks pretty narrow, the region tracks the largest dimensions of the map, which gives it a surprisingly large total area.

The situation is marginally worse with road atlases, as they also have awkward bits to either side of the central crease. Doing the same geometrical sums, it turns out that for a typical atlas page, there's around a fifty per cent chance of a given location being inconveniently positioned on the page.

Cartographers have been trying to combat Murphy's Law of Maps for years and have introduced innovations such as fold-out flaps on one edge of the map. These alter the geometry of the pages and thus the relative areas of map and awkward bits, though not by much. Fortunately, in 2002, that wonderful British institution the Ordnance Survey introduced the ultimate solution to Murphy's Law of Maps, the OS Select service, which produces customized maps centred on anywhere in Britain.

? When driving, why do we arrive at obstacles like narrow bridges at the same moment as a car coming the other way?

Many scientists would probably dismiss this as mere selective memory: that is, people just forget the times when they encounter such obstacles and drive past unhindered. While this may well be part of the explanation, I suspect there is something more going on – tied to the fact that, on seeing an obstacle ahead, we look to see what cars coming the other way are likely to cause problems. Clearly, cars much closer than us to the obstacle, or much further away, aren't likely to be a problem, so very sensibly we fret most about those around the same distance away. Given that cars on a stretch of road tend to travel at around the same speed, this makes it a racing certainty we'll reach the obstacle at the same moment as the car coming the other way. So why are we so surprised? Probably because, sitting in our own car, we only see cars speeding towards us – and tend not to realize that we're both moving towards each other at similar speeds.

? Why does the outward part of a journey seem longer than the return?

This seems to be a very common perception and in my experience, the effect is strongest when making a journey for the first time. I am not aware of any formal research but the most plausible explanation I've come across lies in the fact that the outward journey ends only when one has arrived right outside the unfamiliar destination. By contrast, the return journey feels as if it's coming to an end once one begins to see familiar landmarks and this can happen when one is still some distance from home. Other factors may also be important. For example, there is the dreaded "Are We There Yet?" syndrome, by which young children make even the shortest journey interminable. This may well be a consequence of the fact that an hour constitutes a ten-fold bigger proportion of a three-year-old's life than of the harassed parent doing the driving – even a 30-year-old could be excused for asking if arrival was imminent after sitting in a car for ten hours.

? Should the milk be put in before or after the tea?

Some say that "milk first" is a tradition based on the need to protect thin china cups from the thermal shock of hot water, while chemists have argued that milk-first combats the production of bitter compounds as the tea brews. British Standard BS 6008: "Method for Preparation of a Liquor of Tea" calls for milk to be put in first, in the ratio of 1.75 ml of milk for every 100 ml of tea, which, by my reckoning, is barely a teaspoonful per cup. I daresay this is all true but it seems only wise to match the amount of milk to the strength of the brewed tea – which can only be judged once you've seen it poured into the cup. I therefore count myself among the milk-after camp, along with George Orwell, who in 1946 wrote an article on tea in the London *Evening Standard* stating: "by putting the tea in first and then stirring as one pours, one can exactly regulate the amount of milk".

? Is there any real benefit in warming the pot before making tea?

Many people seem convinced that swilling boiling water around the pot and then pouring it away is a key part of making the perfect cup of tea, the idea being that it helps the pot stay hotter longer, ensuring a better brew. Given the thermal inertia of the average teapot, I doubt this ritual makes much difference. More likely, it removes any of the bitter-tasting tannins remaining in the pot from the previous brew.

? Why does a kettle of water quieten down just before it starts to boil?

The "lull before the steam" effect is linked to the way the water is heated. Kettles are usually warmed from their base, so that the water down there reaches boiling point first. As the bubbles of vapour form, they rise up through cooler liquid, cooling as they go. Unable to maintain enough pressure to keep the surrounding liquid at bay, they suddenly collapse with a pop. Put enough of those together and you get the familiar rumble of a kettle heating up. As the heating continues, however, the bulk of the liquid starts to reach boiling temperatures, allowing big bubbles to go the distance, right to the surface. The result is a deeper, quieter sound – signalling the onset of boiling throughout the liquid, and tea time.

? Is it true that boiling water makes ice cubes more quickly than cold?

Over the years, I've come across several twists on this one, such as "do very hot cups of tea reach a drinkable temperature faster than those that are merely hot?". All involve the phenomenon of cooling but include some traps for the unwary. The basic rule is that the bigger the temperature difference between an object and its surroundings, the faster it cools. So, a very hot cup of tea will indeed cool more quickly than a less hot one. That doesn't mean it becomes drinkable sooner: it just means that it quickly reaches the same temperature as the other cup – but then cools at just the same rate. Add in the fact that it was much hotter to start with and it's clear that the hotter cup will actually take *longer* to become drinkable. By the same logic, it's also clear that boiling water cannot freeze faster than cold water. Well, up to a point. The trouble is, boiled water isn't just hotter than cold water: it also contains fewer dissolved gases and thus freezes at a slightly higher temperature. Evaporation also reduces the mass of water that has to be chilled. In the right conditions, the result can be faster formation of ice-cubes – a phenomenon sometimes termed the *Mpemba Effect*, after the Tanzanian student who first discovered it in 1969.

❓ Is there a fast way of removing ice from car windscreens?

To escape the chore of scraping with an old credit card, or the expense of special aerosol sprays, many people are tempted to use warm water – only to find their windscreen freezing over again once they are on the road, with potentially lethal consequences. Iced-up windscreens are a sign that temperatures have been below zero for quite some time, so the glass will require a fair amount of heating to warm back up. Pouring hot water will melt the ice but once that's done, the thin layer of tepid, rapidly-evaporating water running down the screen has little heat left to warm the glass and quickly turns back to ice.

Happily, this doesn't mean that we are condemned to using those tedious scrapers. For ice to re-form, we need more than just sub-zero temperatures: pretty obviously, there also has to be water. So, the secret to rapid removal of ice is to pour warm water over the windscreen *and* the windscreen wipers – and then quickly switch the wipers on at their highest speed. The wipers remove the thin layer of water that would otherwise turn into ice, keeping the windscreen ice-free until you are on the road and the windscreen heater is working. For side windows, pour the warm water on from the top and remove the meltwater with a rubber-edged window-cleaner. It's a trick I've used successfully in the Alps at minus 18 °C (just remember to thaw out the rubber of the wipers with the warm water first, to prevent them tearing).

❓ What makes Super Glue so strong?

Super Glue was discovered in 1942 by Dr Harry Coover of the Eastman Kodak Company, during research into materials suitable for making transparent components. Searching for a suitable material for a plastic gun sights, he investigated the properties of the compound methyl cyanoacrylate but found it had an annoying propensity to stick to anything with which it came into contact. Dr Coover came across methyl cyanoacrylate again nine years later, while supervising a team working on heat-resistant plastics for fighter aircraft canopies. Again, the substance proved frustratingly sticky but this time it dawned on Dr Coover that he had discovered a new form of adhesive, one which required neither pressure nor heat to glue objects together. Remarkably, the glue was activated by the presence of even minute amounts of water – such as the layer of moisture that, as a result of natural humidity, coats everything. Kodak took up the product and it was first marketed in 1958.

The strength of Super Glue comes from its ability to turn itself from a collection of individual molecules into a chain whose links are extremely hard to break. Electrons from water molecules affect the bond between two carbon atoms in the basic cyano-acrylate molecule, turning it into a double-ended hook that can link up with other glue molecules. These in turn supply electrons to other molecules, triggering the formation of a chain of molecules that binds objects together with impressive strength. The process is so fast and so sensitive to the presence of water that the manufacturers of Super Glue mix it with a tiny amount of acid, to stop it curing too easily. A little moisture – even that on a finger-tip – is enough to undermine the acid stabilizer and the chain-forming process gets underway with a vengeance.

Having been perhaps a little slow to realize the potential of Super Glue, Dr Coover was well ahead of the game in patenting its use in surgery, gluing human tissue together without the need for stitches. Super Glue was first used in this way in the Vietnam War and is now often used in minor surgery.

? Why is a bicycle more stable once it's moving?

I used to think the answer to this one was a combination of simple mechanics plus practice. That is, when learning to ride a bike we train ourselves to cancel out the wobbling we get on starting off by shifting our weight slightly, via the handlebars. Once we're on our way, the gyroscopic effect of the spinning wheels starts to take effect, making the bike even more stable. However, I've learned that the dynamics of anything involving spinning bits is never straightforward and routinely discuss such questions with my personal guru on theoretical mechanics, Dr Ron Harrison, erstwhile lecturer in the subject at City University, London. After several weeks of reading around the subject, calculations and email exchanges with Dr Harrison, my suspicions that this is a very tough question have been confirmed.

Even those who have studied the problem in depth do not agree on the finer details of bicycle stability. They do at least agree on one thing: bikes don't need riders to stay upright; just a push to get them travelling above around five miles per hour will do the trick. It's the source of the inherent stability that causes the arguments. Detailed analysis shows that the much-vaunted gyroscopic effect of the spinning wheels, often thought to explain bike stability (not least by science columnists), is pretty unimportant. Surprisingly, the chief reason moving bikes stay upright is because of the shape of the forks holding the wheels in place. These usually point forward towards the ground, so that an imaginary line drawn from them would touch the ground some way ahead of the wheel. It may only amount to an inch or so but this amount of "trail" makes all the difference to bike stability, bringing forces to bear on the wheels that damp out wobbles. According to Dr Harrison's calculations, a bike with vertical forks – and thus no "trail" – would wobble ever faster as it speeds up, becoming increasingly hard to control. Of course, with practice, any bike is rideable – even a unicycle, whose vertical forks give it zero trail. Even so, it's clear that those outrageously-angled forks on Peter Fonda's motorbike in *Easy Rider* aren't quite so silly after all.

? Why does the British tax year start in the first week of April?

The odd timing of the tax year is a hangover from the days when the New Year began on 25 March – the Feast of the Annunciation; nine months before Christ's birth-day. That all changed when, in 1752, Britain adopted the Gregorian calendar, a reform which moved New Year's Day to 1 January but left the financial year unchanged. The reform also added extra days on to the old calendar dates, so that the start of the tax year was shunted from the last week of March to the first week of April – where it's stayed ever since.

? Can magnets wear out?

Toy magnets sometimes come with warnings not to drop them or heat them up, lest they lose their magnetic power. Yet, even if they are treated with care, magnets eventually, albeit very slowly, lose their strength. This is because magnets owe their properties to the existence of vast numbers of "domains", each around 1 mm across, packed with atoms whose spinning electrons are aligned. This highly ordered state is the origin of the magnetism and also of the vulnerability of magnets to heating or dropping, which jolts some of the domains out of alignment. Even if a magnet avoids such a fate, it will eventually fall prey to the effects of ambient heat and electromagnetic fields, which damage the alignment and steadily weaken the magnetic force. Fortunately, it is a very slow process; a modern samarium-cobalt magnet takes around 700 years to lose half its strength.

? Why are only some substances magnetic?

For all its familiarity, magnetism is a manifestation of very fundamental physics, ultimately linked to the orbital motion and spin of the electrons in atoms. Certain arrangements of these spins – such as those in iron atoms – create intensely magnetic materials but as all substances contain spinning electrons, everything is magnetic: water, wood – even frogs. The reason we don't think of them as magnetic is because the effect is pretty feeble, around a billion times weaker than in metals like iron. Thus, while it is possible to pick up a toothpick using a magnet, it requires an intense magnetic field – around 200,000 times stronger than the Earth's. It can be done: the German physicist Werner Braunbeck levitated small bits of apparently non-magnetic graphite in 1939, while French physicists did the same with blobs of water in 1991. The most spectacular achievement so far came in 1997, when a team led by Dr Andre Geim at the University of Nijmegen in the Netherlands succeeded in levitating a frog. And yes, humans could also be levitated – at least in principle. However, generating the necessary magnetic field would require the output of a nuclear power station, which seems a bit excessive for a party trick.

? Why does a magnet held near a television produce weird colours?

This baffled scientists for almost half a century. The effect was discovered in 1858, well before the invention of television, by the German physicist Julius Pluecker, who experimented with passing electricity through glass tubes from which virtually all the air had been sucked out. He found that the eerie glow given off by these tubes could be bent by magnets. The explanation emerged in 1897, when the Cambridge physicist Joseph Thomson showed that the glow was caused by what we now call electrons and that these can be affected by magnetic forces. The average television (or computer screen) exploits the phenomenon to create coloured images, firing electrons at the screen to create the picture. When a bar magnet is brought near the screen, the electrons are pulled off course, creating weird colours. Moving the magnet around isn't hazardous to health but it can damage the image, by permanently magnetizing the system used to direct the electron beam. Most modern screens have "degaussing" circuitry, which eliminates any lingering magnetism every time they're switched on but personally I wouldn't take the risk, and would steer any magnet-wielding children away from the television and computer.

❓ Why do television screens have red, blue and green dots?

At school we are taught that to create all the colours of the rainbow, one needs red, blue and yellow but not green. The explan-ation lies in the difference between colours formed by reflected light – for example, from paint – and those formed by emitted light. Paints produce colour by absorbing all but one colour from the white light striking them. This "subtraction" method demands the presence of yellow, as, when this is combined with blue, it mops up everything but green. On the other hand, televisions create colours directly, using emitted light and to produce all colours by this "addition" method requires red, blue and green.

❓ Do microwave cookers destroy vitamins in food?

They certainly could – just as any form of cooking would, if the food were heated to destruction. If used correctly, so that they don't overheat the food, the end result of microwaving will be at least as nutritious as that of conventional cooking: there is nothing special about microwaves that makes them more likely to destroy vitamins or other nutrients. The possibility of damage to nutrients was extensively reviewed by scientists in the early 1980s and the conclusion was that there were no significant harmful effects. Indeed, the only surprising finding was that microwaving at low power can mean the food retains more of its nutritional value than it would if cooked conventionally.

❓ Why does blood look blue under the skin?

There was a time when I thought the blood coursing through our bodies really was blue and only became red when it came into contact with air. Having my first blood test put me right on that – but I still couldn't understand why it looked blue under the skin. The explanation emerged in the mid-1990s, following research by Dr Lothar Lilge and colleagues at the Ontario Laser and Light Wave Research Centre, Canada. They showed that when light strikes white skin, the longer, redder wavelengths penetrate deeper and end up absorbed by the blood vessels. As a result, the light reflected back from the skin over a blood vessel has a relatively high proportion of the shorter, blue-violet, wavelengths – making the blood look as if it really is blue. The effect isn't as obvious with dark-skinned people, as the melanin responsible for their skin tone absorbs almost all the wavelengths of light at the skin's surface.

❓ What causes the interference on the radio when moving around?

These annoying "susch-susch" noises are due to *multipath interference*, caused by the radio picking up reflections of FM signals. Having a wavelength of just a few metres, such signals are vulnerable to reflections from everything from mountains to office blocks to cranes – even people walking by. Moving the radio just a few inches can be enough to cure the problem but if that doesn't work you may need to invest in something a bit more sophisticated than the bits of wire manufacturers usually supply as an aerial.

? Why do small aircraft seem to travel so much faster than airliners?

It's an optical effect called parallax. We judge the speeds of objects by comparing how quickly they pass across our field of view. Small aircraft typically fly at much lower altitudes than airliners, making them closer to us, which means they cross our field of view much more quickly, despite their much lower speed. Similarly, orbiting satellites can, despite travelling at over 18,000 mph, seem to move more slowly across the night sky than aircraft, because they are so far away. Parallax was put to impressive use by Gerry Anderson and his team of animators in *Thunderbirds*, when they created the illusion of huge spaces by having layers of scenery behind model vehicles and moving the layers progressively more slowly the further they were from the models – just as they would appear to do in real life.

? Why do boomerangs come back?

Boomerangs are commonly thought to be the invention of Aboriginal Australians but over the years they have turned up at archaeological sites as far apart as Arizona and India; the oldest known specimen, around 23,000 years old, carved from mammoth tusk, was found in a cave in Oblazowa Rock, southern Poland, in 1987. It thus seems that the spectacular flying abilities of boomerangs have been independently discovered many times. Despite their prehistoric origins, their aerodynamics are anything but primitive and exploit an ingenious combination of lift and spin. To provide the lift, the boomerang is launched from an almost upright position, at an angle of around 15 degrees to the horizontal. This allows it to spin and cut through the air at an oblique angle, deflecting it and generating lift as it goes. The famed ability of the boomerang to return home again comes from the fact that, because of its spin, the upper part of its "wing" cuts through the incoming air at a relatively higher speed than the lower part and thus generates more lift. This creates a turning force on the boomerang, which starts to follow a circular path, rather like a spinning toy gyroscope slowly pirouetting on its stand. Oddly enough, just how big a circular path the boomerang follows doesn't depend on how hard it's thrown, or how fast it spins: it's fixed primarily by its wingspan and shape, which are a compromise between spin and lift. Thrown well, boomerangs will happily fly long, looping trajectories of over 400 ft.

? **What causes the lemon pips in a gin and tonic to rise and fall?**

It's not just lemon pips in gin and tonic; the same weird phenomenon can be seen by dropping a raisin into fresh fizzy water. First, it will sink to the bottom of the glass, then perform an odd little dance, before coming back up to the surface and starting the cycle again. The explanation lies in the effect of the crinkly surface of the pip on the carbon dioxide in the drink. Being slightly denser than the liquid, the pip sinks to the bottom, where the dissolved carbon dioxide molecules become trapped in its nooks and crannies. After a while, enough accumulate to form bubbles, which boost the effective volume of the pip to the point where its density becomes lower than that of the surrounding liquid and it starts to rise. On the surface, the bubbles burst, the gas molecules escape, the pip loses its buoyancy – and it sinks, beginning a repeat performance.

? How high is the pressure inside an aerosol can?

According to the British Aerosol Manufacturers' Association, aerosol cans are pressurized to between 2 and 8 atmospheres, equivalent to around 30 to 120 pounds per square inch. If punctured or heated they can explode with potentially lethal violence, especially if they also contain an inflammable fluid. For that reason, the cans are only part-filled with fluid and designed to withstand some expansion, through their concave base and tops. Aerosol cans are also individually tested, before leaving the factory, by being passed through a hot water bath to raise the pressure in the can and test its strength and integrity. They also carry warnings about the need to avoid using near sources of heat. Sadly, these are not always heeded and the effects can be pretty devastating. In February 2000, an elderly woman in Maryland allowed some aerosol cans in her mobile home to come into contact with the pilot light of her gas stove and they exploded like grenades, blowing out the windows and making the walls buckle. It's worth bearing in mind that it is not just a naked flame that can cause an aerosol can to detonate: in hot summers aerosol cans have been known to explode simply after being left on the seats of cars parked in the sun.

? How do the stripes get into striped toothpaste?

No, the tube isn't filled with stripy toothpaste. Indeed, most of it is filled with plain white toothpaste; the trick to the stripes lies in the top of the tube, whose sloping neck contains coloured gel. When the tube is squeezed, the white paste flows towards the open neck of the tube and on its way pushes up against the sloping part. This causes some of the coloured gel to be squeezed into slots that run around the inside of the nozzle at the very top of the tube, directing it on to the surface of the emerging white paste – resulting in stripy toothpaste.

? How do spectacles with "pinhole" lenses work?

Sometimes called *stenopaeic spectacles* (from the Greek for "holed"), these are often advertised as a remedy for myopia and they look like a complete con. A trawl of the medical literature failed to provide any evidence to back the claims – but neither did it produce any evidence contradicting them.

Whatever; the basic principle makes scientific sense. Blurred vision is caused by the failure of the eye to bring all the rays of light entering the eyeball to a sharp focus at the retina, producing multiple overlapping images, which appear as "blurring". One obvious way to reduce this is to block out all but those rays coming more or less straight into the eye – an effect known to photographers as "stopping down". This is precisely what the stenopaeic spectacles do; the pinholes drilled into the "lens" screen out all but those rays heading straight into the eye. The effect is much greater depth of focus, with objects at a wide range of distances all appearing less blurred.

One drawback is that, as less light enters the eye, everything looks somewhat darker. Far more seriously, the pinholes all but eliminate peripheral vision. Stenopaeic glasses are thus potentially lethal when driving or operating machinery. Even so, they seem perfectly safe for tasks like reading, especially if one can't be bothered – or afford – to buy prescription sunglasses. In emergencies (such as when the football results come up on television) when I can't find my glasses, I have resorted to an even simpler solution and improved my vision by peering through the holes in a Rich Tea biscuit.

? Why is aluminium foil painful if it touches your fillings?

The unpleasant tingling sensation feels like an electric shock – which is pretty much what it is. When a tiny bit of foil from, say, a sweet wrapper hits the mercury amalgam in fillings, it creates a tiny battery, with electrons flowing from the aluminium and into your filling via the saliva in your mouth. The current is very weak – but still big enough to be detected by the nerve endings under your teeth.

? How does a quartz crystal allow clocks to keep such good time?

Invented in 1927, by engineer Warren Marrison at Bell Laboratories in New Jersey, the quartz clock relies on the *piezoelectric effect*. When certain types of crystal – such as quartz – are squeezed or stretched, their atoms produce an electric field. In quartz clocks, the flip side of this effect is used; that is, an electric field applied to the crystal makes it change shape. Marrison realised that by applying an alternating voltage to such crystals, they could be persuaded to vibrate at rates of anything from 33,000 to over 4 million times per second, maintaining that rate with extraordinary precision.

Using electronics and mechanical gearing, Marrison was able to create a quartz clock accurate to one second a decade – a tenfold improvement on the best electrically-powered clocks then available. Cramming all the necessary gubbins into something that could be worn on the wrist took another forty years: the first quartz watches were marketed by the Japanese company Seiko, on Christmas Day 1969.

? Do ice-skates really work by the pressure of the blades melting the ice?

One of the highlights of my school physics education took place one lunch hour, when Mr Naden took a block of ice, draped piano-wire over it and attached huge weights to each end of the wire. Over the course of lunch, the wire magically tunnelled its way towards the centre of the block, leaving no trace of its path of descent. This, I was told, was a demonstration of *regelation*, in which high pressure – produced, in this case, by the weights hanging from the wire – lowers the melting point of ice, after which the ice re-forms, sealing the block back up as the wire descends. Some years later, I heard this same phenomenon put forward as an explanation of why ice-skaters glide so smoothly across the rink, as the pressure exerted by the skater's weight creates a thin film of water under the blades. This seemed reasonable enough, but as Professor Clifford Swartz shows in his excellent *Back of the Envelope Physics* (Johns Hopkins University Press 2003), it always helps to put in the numbers. Taking a 60 kg skater and typical blade dimensions, he showed that the pressure generated would have barely any effect on the melting point of ice.

The correct explanation was put forward by the Victorian physicist Michael Faraday. As water turns to ice, a thin layer of water molecules remains on the surface – it is this that makes the ice slippery, rather than skate pressure. Below around minus 10 °C, however, this layer vanishes, making the ice far less slippery.

? How do gun silencers work?

By allowing the explosive gases from the gunshot to expand into a chamber before hitting the surrounding air, silencers can cut noise by a factor of 10,000, or even more. The trouble is, guns are so noisy in the first place – around 100 decibels (dB) for most guns and as much as 140 dB for a shotgun (enough to cause instant deafness) – that they still make a fair bit of noise, even when fitted with a silencer. However, by using sub-sonic ammunition, specially-designed guns and a state-of-the-art silencer, the effect can be very impressive: essentially, just the noise of the bolt hitting the cartridge, followed by a whirring noise as the bullet leaves the gun.

? Is it true that light bulbs last longer if they are left on?

The very first conspiracy theory I heard centred on the longevity of light bulbs and how manufacturers knew how to make the things last for decades but cynically produced duff ones to keep the cash coming in. It's true that companies could make much longer-lasting lights: energy-efficient ones have ten times the lifetime of conventional bulbs. As manufacturers have found, however, people are reluctant to pay a premium for a long-term gain like increased longevity.

Can we do anything to make the throwaway variety last longer? The Institution of Lighting Engineers points out that the lifetime of a conventional light depends greatly on the voltage applied to it, so a dimmer circuit can help. However, the real killer for filament lights is the dramatic change in temperature caused by switching them on and off, which can be well over 2,000 °C. This "thermal cycling" damages the filament, causing cracks that eventually break it altogether (which is why lights usually go pop when switched on).

The obvious way to avoid thermal cycling and thus increase the lifetime of the light, is to leave the thing on permanently. It seems to have worked very well for the light bulb at the head-quarters of the Livermore-Pleasanton Fire Department in California, which is still going strong after being switched on in 1901. (Admittedly, "going strong" is a bit of an exaggeration: as a night-light, the bulb puts out just 4 watts.)

Just how much longer conventional lights would burn if left on continuously isn't clear; frankly, I doubt it's worth the candle and it's not exactly eco-friendly.

? Why do "energy efficient" lights seem dimmer than incandescent ones?

Although more expensive than their conventional counterparts, energy-efficient lights are supposed to be just as bright but use only around twenty per cent of the energy. They are also supposed to last up to 15,000 hours, well over ten times as long as an ordinary light bulb. Even so, people often suspect that a 60 W energy-efficient light looks distinctly dimmer that the old 60 W unit they swapped it for. The explanation lies in the fundamentally different way the two types of lamp generate light. A conventional light bulb allows electricity to flow into a very thin tungsten filament, which heats up to over 2500 °C, becoming white-hot and very bright. In contrast, low-energy lamps are essentially miniature fluorescent tubes, which produce light by passing electricity into a sealed tube containing very low-pressure gas. The gas responds by emitting ultraviolet (UV) light which, when it strikes a special coating on the inside of the tube, is converted into visible light. According to the Institution of Lighting Engineers the whole process takes some time to get going, so that when a low-energy light is first switched on it emits only about eighty per cent of its ultimate level of light. Thus, anyone switching on one of these lamps will think the light level is worse than that from their old incandescent bulb, for the simple reason that – for thirty seconds or so – it is.

❓ Why does polish make shoes look shiny?

We perceive objects to be shiny if they do a reasonable job of reflecting light back in parallel lines. A smooth, flat surface like a mirror does this very well, while rough surfaces bounce light all over the place. The wax of shoe polish works by filling in all the irregularities caused by scuffing, allowing light to bounce back in a more orderly way.

❓ Can microchips wear out through over-use?

While they might not appear to have any moving parts, microprocessors have to shunt electrons around generating heat that can cause malfunctions. Then there is damage caused by static electricity, natural radiation and the build up of oxide inside its wafer-like structure. The upshot is that microprocessors in home computers last for around ten to fifteen years – which for many of us amounts to indefinitely, as we've usually been forced to upgrade our computer to cope with the latest bloated software long before then.

? Why is it easier to mop up water with a wet cloth, rather than a dry one?

To do a decent job of mopping up we need a force strong enough to counteract gravity and lift the water off the floor. With a dry cloth, that comes from the molecular forces between the material making up the nooks and crannies of the cloth and the water molecules (what is often called "capillary action") driving water up into the cloth. With a wet cloth, the water molecules already in the cloth can exert their relatively strong inter-molecular force on the water still on the ground, mopping it up much more effectively.

? What is the fastest way of finding someone who has wandered off?

The obvious method is to arrange beforehand that, if lost, the person will go immediately to a pre-arranged meeting point and tell staff what has happened, so that an announcement can be made. Yet, as anyone with children (or dogs) will know, the best-laid schemes go awry, so it is as well to have Plan B. I can personally recommend a method that has its origins during the Second World War in submarine hunting. Devised by Professor Lyn Thomas of Southampton University, it focuses on the fact that the longer the time spent searching, the greater the chance of success – but the greater the risk of wasting time continuing to look for someone who has already been found. The trick is for the searchers to agree on an initial time limit – say fifteen minutes – during which they will hunt for the person before meeting to compare notes. If the initial search is unsuccessful, then another search begins but for a shorter time – say twelve minutes – and so on. This makes best use of the search time available and from personal experience works extremely well, especially if the rendezvous point is well-chosen. Some time ago I lost the family dog on a walk and spent a fruitless half hour scouring the park for him. As I walked to the car-park wondering what to tell the children, there was Roly, waiting next to the car – which, unlike me, he had realized would make an obvious rendezvous point. Dog: 1; Human: Nil.

? How is the non-stick coating stuck on to kitchenware?

Discovered in 1938, by the American chemist Dr Roy Plunkett of the DuPont Company, polytetrafluoroethylene (PTFE) – Teflon, to use its commercial name – is famed for its non-stick properties. It is also incredibly stable and non-reactive, resisting acids, alkalis, heat and solvents. All this is wonderful until one comes to actually using the stuff as a coating, when its chemical stubbornness becomes a liability. The problem of getting PTFE to stick to metal surfaces was solved in 1954, by Louis Hartmann. He used acid to etch tiny holes in the surface of the metal, then applied a layer of PTFE and baked it at 400°C. The melted PTFE seeped into the holes and solidified, pinning the coating to the surface. The date of Hartmann's work is significant: fifteen years before Armstrong and Aldrin stepped on to the moon – giving the lie to that old saw about the only useful thing to come out of the Apollo programme being the non-stick frying pan.

? What are the "protective atmospheres" that supermarket food is now packaged in?

Much as consumers deplore wasted food, every supermarket manager knows that we studiously avoid choosing produce that looks anything less than perfect. The backlash against preservatives has compelled food technologists to find ever more sophisticated ways of allowing produce to look fresh for longer, including the use of protective atmospheres. These are based on the biochemistry responsible for food going "off". For example, the ethylene released during ripening is the cause of discoloration of vegetables like broccoli, while the browning of vegetables has been traced to the action of certain enzymes. Exposing fresh meat to air for too long causes it to turn an unappealing brown, as the oxymyoglobin responsible for giving it a bright red colour turns into metmyoglobin.

During Modified Atmosphere Packaging (MAP), the produce is sealed in a special mixture of gases that slows down the biochemical reactions that cause the trouble. For example, a carbon dioxide-rich MAP slows bacterial spoilage, boosts the shelf life of vegetables by up to five times and allows meat to retain its looks for at least forty-eight hours. Nitrogen-based MAPs slow enzymatic browning in vegetables, while a 60/40 mix of carbon dioxide and nitrogen is used to preserve fatty fish like herring and mackerel.

The transparent film in which produce is wrapped is also modified to slow food spoilage; "ethylene-scavenging" films are used to mop up the gas that causes discoloration in green vegetables.

Apparently, the current big challenge is to find ways of slowing the rate at which bread goes off. The preferred solution in our house is the technical process called "eating it". It is, however, not without side-effects.

? What causes the electric shock one gets on leaving a car – and how can it be avoided?

A car can become charged with static electricity by the frictional effect of the air rushing over it but the principal cause is the very act of getting out of the car. As you slide over the seat, the friction strips electrons out of the fibres of clothes, producing a hefty electric charge. Given half a chance – like when you put the key into the metal car door to lock it – this charge will flow to earth and if the route is via your finger, the result can be, well, shocking. It can also be potentially lethal if the discharge occurs at a petrol station: such sparks have been known to trigger serious fires. There have been some attempts to find materials that reduce the charging effect, but the physics of static electricity is still not well-understood and the results are unimpressive. Some people suggest avoiding the static build up by using those awful car seat covers made from wooden balls but a technique I've found works well is to spread out your hand and swiftly touch the car with the base of the palm. The density of charge around this relatively smooth, round part of the hand is much lower than it is at the fingertips, so the discharge is imperceptible. The same trick works in shops whose nylon carpets produce fearsome discharges when one touches any metal. Again, touching the base of the palm to any metal connected to earth, like a clothes rail, will prevent a nasty shock.

? Why does metal spark when placed in a microwave oven but other substances do not?

Microwave ovens generate waves of electromagnetic energy capable of stripping electrons from the atoms of substances put inside them. The electrons in metals are especially mobile (which is why metals conduct heat and electricity so well) and tend to build up around any sharp points (such as the tines of forks) left in the oven. If the microwave field is strong enough, these electrons smash into the surrounding air so violently that they can flow through it – triggering a mini lightning bolt. Metals don't always cause sparks, though: if there's no point on the object where the microwave field strength reaches the critical level needed for the air to break down, then there won't be any sparks (or "arcing" as it's called in the trade). I recently bought a metal non-stick dish that works perfectly well inside a microwave oven and cooks excellent bacon and eggs in around six minutes flat.

? Do air fresheners do more than merely mask smells?

As well as masking scents, like pine and lemon, commercial air fresheners include compounds called malodour counteractants, which specifically block unpleasant smells. Discovered by scientists at Monsanto in the mid-1970s, exactly how they work is unclear. According to Professor Tim Jacob, an expert on scent perception at Cardiff University, these volatile organic chemicals may act like an anaesthetic, preventing the scent receptors in our nose from reacting to substances such as sulphur, often found in foul smells. Alternatively, they may block the transmission of the signal back to the brain; no one knows.

❓ Why is ketchup so difficult to pour from a new bottle?

This is the phenomenon of *thixotropy*, which ketchup shares with quicksand and non-drip paint. Left alone, these fluids are relatively thick and sticky but become much runnier when exposed to a sliding force – like, say, a brush dragged across a tray of non-drip paint. All are chemical mixtures whose bonds behave like a house of cards: they can withstand some downward pressure but fall apart when subjected to a sliding, or "shear" force. The trouble with new ketchup bottles is that they're full, so there is little exposed surface area from which the ketchup can be freed by the shear forces created by shaking. As the bottle empties, it becomes easier to get the ketchup out – especially if the bottle is held at an angle, to expose as much surface area as possible. Or, as the American poet Richard Armour put it "Shake and shake the ketchup bottle. None'll come, and then a lot'll".

❓ Why does aftershave always feel cold on the skin?

Aftershaves and perfumes typically have an alcohol content of well over eighty per cent; this alcohol is the solvent for the oils that give the scent. As well as being an excellent solvent, alcohol is extremely volatile, meaning that its molecules take little persuading to evaporate. As they leave your skin, they extract heat energy on the way – giving the familiar chill on even a hot day.

? Why do the bubbles in a glass of Guinness sink instead of rise?

As bubbles are less dense than the surrounding liquid, one might reasonably expect that Newton's law of gravity would apply in the glass of beer and the bubbles would rise. Clearly some other force is predominant, the nature of which was uncovered recently by Professor Clive Fletcher and his colleagues at the University of New South Wales in Sydney. Computer modelling showed that relatively large bubbles, hidden from view in the middle of the beer, are indeed rising and dragging liquid up with them. At the surface, the beer is released and starts to descend close to the walls of the glass, dragging tiny bubbles down with it. The reason stout shows the effect where a thin and weedy lager doesn't is largely due to its relatively high "stickiness", which allows the bubbles and the liquid to interact so effectively.

? What causes microwaved milk to explode out of a mug?

Microwave ovens are notorious for uneven heating, creating "hot spots" in food. In liquids, this can create pockets of unstable superheated liquid at temperatures above the normal boiling point. Adding just a few grains of sugar can be enough to allow the superheated vapour to gather around them, forming rapidly-expanding bubbles, that causes the overlying liquid to erupt. The reason it happens only rarely is because most cups have tiny cracks or specks of dust in them which allow bubbles to form more gently during heating. This makes brand-new cups and ones taken straight out of the dishwasher particularly prone to the Vesuvius Effect. To minimize the risk of being scalded (and dozens of people are hospitalized because of it every year), make sure the heating time is reasonable (no more than a couple of minutes or so for a mug of milk), stir it after a minute or so and leave it to cool for a while after taking it out.

? Why is there so often a strong wind around tower blocks?

Even on a calm day, there is usually a breeze blowing around the base of tall buildings, sending litter swirling high into the sky. Watching these bits of paper gives a clue to what is going on: they often fly round and round, trapped in vortices around the base of the buildings. This is the tell-tale sign of the buildings acting like giant sails, directing the relatively strong high-level winds that blow, unfelt, over our heads, down to the ground, where they strike with considerable force. It is not just the strength of the winds that causes problems: the downpour of cold air also robs the buildings of heat. Unfortunately, this passed over the heads of 1960s architects, who built communal playgrounds around the base of tower blocks, which were duly shunned by children keen to avoid flying objects and frostbite.

? Can a mirror reflect things that are not immediately in front of it?

Indeed it can, as is neatly demonstrated by the half-height mirrors found in many changing rooms. Despite only extending from head-height down to the waist, it is still possible to see our full length – a consequence of the fact that light striking a mirror at an angle bounces off it at the same angle.

? Why do bathroom scales give different readings on hard surfaces and carpets?

Dieters have long insisted that their weight can vary enormously, depending on where they put their bathroom scales. According to reports, it was possible to lose several kilograms simply by putting the scales on a bare floor rather than carpet. Until recently, it was unclear why bathroom carpet exerted this extra gravitational attraction – or which reading to believe.

The answers have now been supplied by Jon Pendergast and Dr David MacKay, of Cambridge University. Examining the mechanism of conventional mechanical bathroom scales, they found that the weight-measuring system relies on a set of pivots in contact with the base-plate of the scales. On a hard surface, this base-plate stands clear of the floor and flexes when a person steps on to the scales. An arrangement of levers then converts this into a reading. On soft carpet, however, the scales sink into the pile and the base-plate cannot flex properly. The result is a faulty reading: the researchers found the error can be as much as ten per cent higher than the true value. So, it seems that many people who think they have a weight problem in fact have a carpet problem. Those who want to retain their bathroom carpeting have two choices: invest in a set of digital scales (which rely on a different mechanism), or kid themselves that they would already be at their ideal weight, were it not for the pernicious effect of their carpet.

? How does soap work?

A bar of soap is a chemical miracle – a little scented block that unleashes molecular forces when dunked in water. Soap molecules have the peculiar property of having one end attracted by water and the other repelled by it. This gives soap two key properties for getting things clean: first, it reduces the attraction that water molecules have for each other, compelling them to spread themselves more effectively over whatever has been put in the soapy water; second, it allows the soap molecules to work their way under dirt and prise it off, floating it away surrounded by bundles of molecules – which also stops the dirt simply floating back on to the clothes again. Oddly enough, although the Babylonians first made soap (presumably by accident) around 4,800 years ago, it was primarily used to treat skin conditions; its dirt-shifting powers weren't recognized until medieval times. Given that so many bacterial and viral infections (especially colds) are transmitted by skin contact, the invention of soap must rate as one of the major medical break-throughs in history.

Its effect on water molecules comes in handy for clearing steamed-up mirrors, when the myriad tiny drops of condensed moisture prevent it reflecting well. A very thin film of soap spread over the mirror breaks the surface tension of the droplets, allowing them to spread out in a nice, flat layer – and gives you back your reflection. Presuming you want it, of course.

? Why do onions make your eyes sting – and what can we do about it?

Slicing or peeling an onion sets off a chain of chemical reactions, the end result of which is tears streaming down our face. The onion cells we have damaged release two types of chemical: organic molecules, amino acid sulphoxides, which give onions their taste and enzymes, allinases. The enzymes turn the sulphoxides into a third compound, syn-propanethial-S-oxide – which is what stings our eyes. In self-defence, they respond by pumping out tears, diluting the concentration of the chemical. Or at least, they do as long as we don't keep wiping them away with fingers covered in the compounds. Various remedies have been put forward over the years; my preferred method is to chop the onions in a bowl of water, which prevents the vapours reaching your eyes. Some people insist that sucking on a sugar cube, lemon or even a piece of bread helps. Quite why, I have no idea.

? What's the secret of those table lamps which come on when touched?

I have long been grateful for the invention of the touch-lamp, which, as I doze off in bed, saves me the effort of finding the switch without singeing my fingers or knocking the whole thing over. They also elegantly resolve what one might call the

Light Switch Paradox: namely, that in a dark room you need to put the lights on in order to find the switch to put the lights on. With a touch-lamp you only need a vague idea of where it might be; touching it anywhere brings light shining forth. They work by having circuitry that puts a small electric charge over the surface of the lamp. Left to itself, the flow of charge in and out of the lamp stays at a certain level. Touch the lamp and it finds itself putting electric charge into something with far larger capacity – you – and the difference is detected by the circuitry. Simple touch-lamps go into the opposite state: "on" if originally "off" and vice versa. Touch-lamps with dimmers have extra circuitry which controls the frequency with which the filament receives current: 33% of the time for the low setting, 66% for the brighter setting and 100% for full blast.

? • Why does liquid water expand below 4 °C, when nearly all other materials continue to contract?

This "anomalous" expansion of water causes pipes to burst in winter, as the ice within them takes up around nine per cent more volume than the original liquid. Like most liquids, water contracts when cooled but at around 4 °C it starts to expand, at first gradually and then very close to 0 °C, at a prodigious rate. Below freezing, the ice shrinks like ordinary solids again. Anomalous expansion is not the only curious feature of water: it is also surprisingly difficult to warm up (needing ten times more energy than an equivalent lump of iron) and has a surface tension higher than glycerol. The cause of all these anomalies lies in the feeble hydrogen bonds within the V-shaped H_2O molecule, which allow water molecules to form short-lived structures in the liquid state, the most common being a 3D arrangement of six water molecules. While the full details have yet to be worked out, it is the behaviour of these hexamer structures that are ultimately responsible for water's many bizarre properties.

Chapter 2

Matters of life and death

? **Do painkillers really work, or do headaches usually disappear of their own accord?**

I often suspect that a headache that disappears a couple of hours after taking aspirin would probably have gone anyway but, lacking access to a parallel universe to see if this is true, I've continued to keep faith in the wonders of medicine. After researching this question, though, I think I may have been a bit too credulous. Over the years, many clinical trials comparing painkillers with placebo have been carried out and the results have been turned into a "league table" by Dr Andrew Moore and his colleagues at the Pain Research Unit of the Churchill Hospital, Oxford. Instead of the usual, more or less incomprehensible, measures adopted by statisticians, the Oxford team stated their results in terms of "Numbers Needed to Treat" (NNT): that is, the number of people who would have to take the medicine in order for just one to benefit. A truly effective drug has an NNT of around 1 – that is, the medicine benefits every one who takes it. Less effective drugs have a higher NNT. The dismal fact is that, for moderate to severe pain, the remedies we can get over the counter all have an NNT of more than 2: in other words, if you take them, the odds are against you feeling a lot better because of them. For example, a couple of 500 mg tablets of paracetamol have an NNT of around 4, while those "new-power-to-relieve-your misery" combinations, like aspirin plus codeine, limp in with an NNT of over 5. Ibuprofen tends to be somewhat better, with an NNT of around 2.5, implying that around forty per cent of people taking it will get substantial relief. Even if not substantial, you may get some relief and it's always possible you are one of the people who responds really well to these drugs. Just don't be surprised if your tense, nervous headache hasn't vanished with the alacrity suggested by the adverts.

? Does being cold and wet increase the chances of catching a cold?

Despite their name (first coined in 1537, in the state papers of King Henry VIII), there seems not to be a direct link between colds and exposure to low temperatures. Studies have been carried out into whether being cold and wet reduces immunity to cold viruses, but the results have been negative. That said, there might be an indirect link: when it's cold outside we are more likely to spend time indoors, in all too close proximity to the viruses of other people. Professor Ronald Eccles, director of the Common Cold Centre at Cardiff University, has suggested that being cold and wet may also lead to a heat-conserving narrowing of the blood vessels in the nose. It's also possible that our disease-fighting immune system is compromized through the psychosomatic effect of being so miserable when thoroughly cold and wet.

? Can zinc tablets really cure the common cold?

For years scientists have insisted that there can be no cure for the common cold, essentially because the viruses responsible mutate so rapidly that a drug developed for any one strain quickly becomes ineffective. There is a little caveat: molecular "canyons" on the surface of cold viruses are less affected by mutations and also play a key role in attacking healthy cells. A compound that blocks off these regions may thus stop a cold in its tracks. One such is zinc gluconate; over the years various research teams have claimed to show that lozenges containing this compound can significantly reduce the duration of colds or even stop them developing at all. The lozenges have been available in high street chemists for some time and people have told me they do seem to work quite well. The trouble with such anecdotes, of course, is that there's no telling how long the cold would have lasted in any case.

? How long does it take to sober up after a few drinks?

Longer than most people think; indeed, it is all too easy to have a few in the evening and still be over the legal drink-drive limit the following morning. There are various ways of estimating the effect of time on how drunk one feels – or, more precisely, on the blood alcohol concentration (BAC) – but the basic results are these. It will take an average 70 kg man around 3 hours to sober up completely from one pint of average strength (3.5%) beer. At that rate, anyone of that weight downing five pints in an evening won't be completely free of alcohol for fifteen hours – well into the following day. They could also still be over the legal drink-drive BAC limit as they set off for work the next morning. As BAC depends on the volume of alcohol in the blood, weight has a major impact on how long it takes to recover – roughly speaking because big people have more body fluid with which to dilute the alcohol. So, a 90 kg man sobers up around twenty-five per cent faster than his 70 kg counterpart.

For women the situation is rather, well, sobering. First, they typically don't weigh as much and so have considerably less body fluid with which to dilute their drink. Second, they also have lower levels of the enzyme ADH, which breaks down alcohol. The upshot is that an average 55 kg woman will take around four hours to sober up completely from a single pint of beer (equivalent to two glasses of wine).

This difference in weight and biochemistry between the sexes also has some interesting implications for women who insist they can match any man's drinking capacity. This is entirely possible – if the woman weighs around fifteen per cent more than the man. So, if a 55 kg woman wants to impress her new 70 kg boyfriend by showing him her capacity for drink, she will also need to impress him with her capacity for food – and put on around 20 kg.

? Could mobile phones cause artificial joints to become painful?

A colleague told me of his suspicions to this effect, having had pain from his artificial joint since he began carrying his mobile in his right trouser pocket, right next to it. He said that, while reluctant to join those who see mobiles as the cause of every medical ailment, he still thinks there may be a link. This struck me as perfectly reasonable, as mobiles emit microwaves even when they are not being used to make calls, in order to signal their whereabouts to the nearest base station. Dr Michael Clark of the National Radiological Protection Board confirmed that it's possible this could induce a tiny stimulus in the nerves around the joint. If so, carrying the mobile in another pocket should solve the problem – though Dr Clark suggests that anyone with persistent symptoms like these should seek medical advice.

? Does homeopathy work?

The ideas behind homeopathy (from the Greek for "same suffering") date back over 2,500 years to the Greek physician Hippocrates, who argued that "By similar substances a disease arises and by administering similar things they regain their health from sickness". This sounds barking mad until one realizes that it is the principle behind vaccination. Given the success of that, it is hardly surprising that some physicians – notably, the eighteenth century German physician Samuel Hahnemann – wondered if the same idea might work for non-infectious ailments. Hahnemann believed there was some empirical support for the idea, in the form of quinine, a fever-producing extract from the bark of a Peruvian tree which was known to be effective against malarial fever.

However, Hahnemann ran into the same problem that had blighted early forms of vaccination: too high a dose was a bad thing. He therefore began diluting and shaking the extracts and in the process put homeopathy into conflict with conventional science, as he claimed that the greater the dilution, the more potent the remedy. This finding is more than merely paradoxical: in many cases, the level of dilution was such that not a single molecule of the original extract remained.

While the scientific community emphasizes the implausibility of the dilution effect and bemoans the lack of any obvious explanation, the public seems interested in nothing but the question: does homeopathy work?

I am with the public on this one. To this day, the mechanism of anaesthesia remains poorly understood but I have yet to hear of scientists declining its use before major surgery. That said, we are surely right to demand more than merely anecdotal evidence for the effectiveness of homeopathy. There is no shortage of studies of homeopathy: hundreds are reported in the medical literature. Sadly, their quality is typically so awful that it is all but impossible to draw firm conclusions. Positive findings are routinely undermined by the fact that the trials weren't "double- blind" (where neither the patients nor the doctors treating them knew who was getting the remedy and who the placebo), with all the risk of self-deception that entails. Negative studies, on the other hand, are often so small they have little chance of detecting anything but the most dramatic levels of effectiveness. In March 2003, a major review in the *Annals of Internal Medicine* concluded that the evidence that homeopathy works is far from compelling for most conditions and that it should not be used if there is a proven alternative. Even so, the researchers considered that for some ailments, notably allergies and childhood diarrhoea, there is now some evidence for its effectiveness. Given the popularity of homeopathy, it is surely time both advocates and sceptics got round a table, agreed on a study design and conducted some decent, rigorous, large-scale clinical trials.

? How do antibiotics kill bacteria?

Most antibiotics are based on the chemicals evolved by bacteria and fungi to attack other organisms: penicillin and its derivatives come from the *Penicillium* fungus, while the tetracyclines are produced by the *Streptomyces* bacterium. Antibiotics attack bacteria in a variety of ways. Some, including penicillin, interfere with proteins linked to bacterial cell walls, making them leaky. Others interfere with vital protein manufacture inside the bacterium, or attack at an even more fundamental level, such as preventing the processing of genes. Natural selection has led to some bacteria acquiring mutations that allow them to resist attack by some antibiotics. For example, the much-feared *Staphylococcus* bacteria acquired a penicillin-blocking enzyme within just four years of the introduction of this "miracle cure".

? Why are epileptic seizures more common during a full moon?

The link between seizures and phases of the moon has a long history and every so often another study claiming to have substantiated the link makes headlines. What one doesn't hear about, of course, are all the studies failing to find such a link and the current consensus is that it's a myth.

❓ What medical condition did the Elephant Man have?

The Elephant Man – real name Joseph Merrick – was born in Leicester in 1862 and the first symptoms of his medical condition emerged while he was still a toddler. By the time he was twelve the growths and bone deformations that gave him his name had set him on a downward spiral that led him to appear in an East End freak show. His "manager", Tom Norman, claimed that the affliction followed an accident in which Merrick's mother was knocked down by a runaway elephant while pregnant. Merrick himself accepted this explanation but it failed to convince the eminent Victorian physician Sir Frederick Treves, who rescued him from the freak show in 1886. Even so, neither Treves nor his contemporaries were able to suggest anything more specific than a kind of nerve-related disease.

Years later, it was suggested that Merrick may have had a severe form of the genetic disorder neurofibromatosis-1 (NF-1), while in 1986 scientists in Canada made a case for it being Proteus Syndrome, an extremely rare congenital tissue-distorting disease identified in the late 1970s.

Neither diagnosis fully fits the facts, however, which prompted Paul Spiring, a biologist based in The Netherlands, to suggest that perhaps Merrick suffered from both conditions. As well as accounting for most of his symptoms, this would explain why nothing like Merrick's case has been seen since. NF-1 occurs in around 1 in 2,500 births, while Proteus Syndrome affects around 1 in 9 million – which, assuming they are unrelated conditions, implies that there should be only one case of combined NF-1/Proteus Syndrome every 22,500 million births. At that rate, it's unlikely that there will be another Elephant Man for several centuries.

❓ What causes "stitch" when running?

One of my abiding childhood memories is of tramping round muddy fields on Saturdays as part of my school's cross-country team, trying to deal with this odd ache with the even odder name (it's actually Anglo-Saxon). Medical scientists today refer to it as Exercise-related Transient Abdominal Pain (ETAP), which sounds very impressive but explains precisely nothing – which is not unrelated to the fact that the cause of stitch remains something of a mystery. Medical textbooks tend to attribute the pain to visceral ischaemia – that is, inadequate blood supply to the muscles of the chest and abdomen during exercise – but what evidence there is points instead to the stretching of ligaments in this area as a more plausible explanation. A recent study showed that stitch most commonly affects swimmers and runners, suggesting a link with repetitive torso movement. Researchers at the University of Otago showed that drinking 1.5 pints of fluid before running triggered stitch, apparently through the weight of the fluid pulling on visceral ligaments. Happily for long-suffering runners, the researchers also found some ways of curing it: bending forward while tightening abdominal muscles, tightening a belt around the waist or breathing deeply through pursed lips. I find that not running works for me.

? Whatever happened to irradiated food?

The idea of using radiation to kill off harmful microbes and lengthen shelf-life has been around for about 100 years and is now approved by forty countries for use with foods ranging from fruit and cereals to shellfish and poultry. Hundreds of studies of the effect of irradiation on food have been carried out and the results have prompted the likes of the World Health Organization to declare that the process is both effective and free of toxic side-effects. In the UK, current legislation permits the irradiation of seven categories of food, yet you'd be hard-pressed to find any examples on UK supermarket shelves, because no company operating in the UK holds the licence needed to irradiate foods other than herbs and spices. Presumably, commercial operators have decided that the facts about irradiated food are neither here nor there: consumers just won't touch food that carries a label linking it to Chernobyl and Hiroshima. It doesn't have to be that way, though. Recent studies have shown that consumers are perfectly happy to buy irradiated food when told of the relative risks and benefits. A few years ago a small food retailer in Chicago sold irradiated and ordinary fruit side by side, together with clear labelling and extensive literature; the irradiated fruit out-sold the ordinary variety by 9 to 1.

? Why can't AIDS be transmitted by mosquitos?

As mosquitos undoubtedly transmit several lethal viral diseases, such as yellow fever and dengue, the idea that they might play a role in spreading HIV is frighteningly plausible. It was a genuine concern among scientists in the very early days of the AIDS epidemic, with stories emerging about mosquitos being responsible for the relatively high prevalence of AIDS in a community in south Florida. Fortunately, laboratory studies have shown that any HIV picked up in the blood meal of a mosquito is destroyed by the insect's digestive enzymes and so not passed through to the mosquito's salivary glands for injection into its next victim. This explains why, over twenty years into the epidemic, the spread of the disease shows no signs of transmission by mosquitos or indeed any other insect.

? Why does mixing drinks cause a bad hangover?

Hangovers are the result of a really nasty chemical double-whammy. First, alcohol interferes with the release of the anti-diuretic hormone vasopressin, which leads to rather a lot of trips to the loo. For every glass of wine, one can expect to lose two to three times the same volume of water. That leads to dehydration, which boosts the concentrations of the toxins contained in the drink along with the pure alcohol (ethanol). These toxins include "congeners" like methanol, which are produced during manufacture and give a drink its characteristic taste, smell, appearance and potential for an agonising hangover. Roughly speaking, the darker the drink, the more congeners it contains and the greater the pain from over-indulgence. Top of the league is cognac, followed by red wine, rum, whisky, white wine, gin and vodka. There is a lot of truth in the old adage about not mixing grape and grain: any one of these dark drinks is bad enough on its own; mixing them leads to cross-reactions and more pain. As for hangover cures, drinking water along with the booze helps enormously. After the event, painkillers can help, though they're best avoided if you feel nauseous, as they can cause more stomach irritation. Some scientists have also claimed that vitamin B6 can also help. In the end, the only proven hangover cure is time.

? Is there any evidence that vegetarians are healthier than omnivores?

Scientists have a habit of responding to such blunt questions with lectures about how it's all far more complicated than one could possibly imagine. In matters of diet, they have a point. First, one must distinguish between the fish-is-ok veggies and fundamentalist vegans on the one hand and burger-munching lard-mountains and occasional chicken-nibblers on the other. The epidemiological evidence is surprisingly consistent: vegetarians do tend to live longer than omnivores, chiefly because they have lower rates of several forms of cancer and heart disease.

The level of benefit is impressive. A twelve year study of almost 2,000 vegetarians by the German Cancer Research Centre in Heidelberg found that the death rate from all causes was half that of the general population, with cancer and heart disease rates being cut by a half and two-thirds respectively.

As soon as one asks why this should be, the scientific hand-wringing and caveats become especially important. Vegetarians live generally healthier lives in any case: they are less likely than omnivores to smoke, get roaring drunk or sit about all day. Lamentably, comparative studies of vegetarians and meat-eaters have thus far been unable to show whether such lifestyle factors are more important than the diet alone.

There are certainly good biochemical reasons for believing that the vegetarian diet is, at least partly, responsible. Analysis of the diets of well-nourished vegetarians show that they have a lower intake of cholesterol, whose artery-clogging effect has been linked to cardiovascular disease. The veggie diet also involves a higher intake of antioxidants, which can protect cells from genetic damage that can lead to cancer. However, hard evidence to support this reasonable-sounding theory has yet to emerge.

As a result, the suspicion that socioeconomic factors account for a substantial part of the apparent health benefits of a vegetarian diet remains. Until conclusive evidence emerges, I for one will continue to follow Kingsley Amis's dictum: there are no pleasures in life worth forgoing in return for two more years in a nursing home in Weston super Mare.

? Why do women typically live longer than men?

Recent research suggests that women retain a strong disease-fighting immune system longer than men. The most widely-accepted explanation, however, is that men are simply exposed to more life-threatening risks – at least, in countries where childbirth is now relatively safe.

? Is counselling really beneficial after a traumatic event?

Reports of major tragedies affecting communities are now routinely ended by a reporter gloomily intoning "Those affected are being offered counselling". If the extant evidence is anything to go by, those affected should turn the offer down. In

2002, *The Lancet* carried a report by researchers at the University of Amsterdam, who had trawled the scientific literature for studies of how people fared after being offered "single session" counselling, in which those affected are advised of the psychological effects they can expect over the coming days. The researchers failed to find any evidence that such advance warning was beneficial. In 2003, Dr Suzanna Rose and her colleagues at the Berkshire Traumatic Stress Service in Reading reviewed the results of randomized controlled trials of such counselling (such trials are generally regarded as the most reliable way of finding out what really works and what does not). They found that three studies pointed to some benefit, six to no benefit and two suggested the counselling had made things worse. Significantly, these latter studies were also the ones which observed the people involved for the longest time.

As always, the researchers called for further research, the suspicion being that counselling leads people to believe they should exhibit the symptoms they have been told to expect. As yet, the mere lack of compelling evidence that they do any good has yet to deter counsellors from rushing in to help.

❓ Does greasy food really cause acne?

Acne is triggered by hormones (principally testosterone) which cause over-production of grease by the skin, leading to blocked-up pores. So, while too much grease is linked to acne, it has nothing to do with grease in the diet. On the contrary, there is a growing suspicion that a high fat diet may actually help, by reducing carbohydrate intake. The idea is that sugar and highly-refined carbohydrates (like bread and cereals) lead to surges of insulin as the body tries to deal with the overload, a process which boosts levels of the hormones responsible for acne.

Support for the link with carbohydrates comes from recent research by a team from Colorado State University, who compared the skin of people living on a low-carb diet in Papua New Guinea and Paraguay with westerners. They found no cases of acne among those on the low-carb diet, yet around half of men and women over the age of twenty-five in the west were affected.

? What is the best way to avoid jet lag?

A friend of mine chalks up 100,000 air miles a year and he swears by melatonin, a hormone, released by the pineal gland, that lets the body know what time of day it is. By taking melatonin for a few days after his flight, he does seem to get his act together with impressive speed. In their review of the literature, published in 2003 in the *British Medical Journal*, Dr Andrew Herxheimer of the Cochrane Centre and Dr Jim Waterhouse of Liverpool John Moores University confirmed that there is now good evidence that taking 2–5 mg of melatonin at bedtime after arrival is effective and it may be worth repeating the dose for the next two to four days.

Unfortunately, while melatonin can be bought in supermarkets in the US, it is not available in the UK without a prescription. This raises the question of what other measures one can take to minimize the impact of jet lag. Herxheimer and Waterhouse point out that the body's internal clock is also influenced by external cues, particularly light levels. After a flight westward they suggest staying awake during daylight, taking a brief nap only if necessary, to ensure one gets most sleep after dark. After an eastward flight, they suggest being awake but avoiding bright light in the morning and being outdoors as much as possible in the afternoon. During any flight, they recommend eating high-fibre foods, like apples and drinking plenty of water and fruit juice rather than tea and coffee.

The duration of the flight and number of time zones crossed are clearly important in determining how awful one will feel on arrival. Eastward journeys seem to be marginally worse than westward, and departure and arrival times can add significantly to jet lag. When presented with a choice of flight, it might therefore be worth bearing in mind that flights that leave between 2200 and 0100 or arrive between 0800 and 1200 are particularly likely to produce jet lag. The best times are departures between 0800 and 1200 and arrivals between 1800 and 2200.

? Why don't gastric fluids digest the stomach?

Given half a chance, they do. The stomach uses a pretty potent form of hydrochloric acid to kill off ingested bacteria and only a thin lining of mucus prevents this from attacking the stomach itself. If this lining is disrupted – most commonly by acid-resistant *Helicobacter pylori* bacteria – the result is irritation, followed by a peptic ulcer and even potentially fatal perforation of the stomach wall.

? What causes the spectacular colours of bruises?

Following a hefty blow to soft tissue, the site of the injury initially turns red, then becomes deep blue and purple. By the end of a week, the first green hues have started to appear, followed by yellows and browns. It can take a fortnight or so for the whole spectacular show to disappear.

The display is the result of a complex series of chemical reactions. The initial red colour is the result of fresh, oxygen-rich, blood leaking from the capillaries that were crushed and broken. After forty-eight hours or so, the failure of this trapped blood to pick up fresh oxygen leads it to darken in colour, making the area purple, blue or even black if the amount of trapped blood is big enough. Over the next few days, the colour changes again to green and yellow as the body sets about breaking down the blood cells. After a week or so, the clean up is well underway and the bruise begins to fade.

? Does reading in poor light damage the eyes?

My first thought was to dismiss this as an example of finger-wagging nonsense handed down through generations of parents. As this says more about my problem with authority than the facts, I spoke to Larry Benjamin, Honorary Secretary of the Royal College of Ophthalmologists. Apparently, there is some evidence from animal experiments (don't ask) that, if eyes are denied light during early life, they are more prone to short-sightedness, though no one knows why. As a child's eyes are still developing in shape until the age of three, it may thus be possible that poor light could lead to defective vision in later life but as few toddlers spend hours reading by torchlight, this is hardly convincing. According to Mr Benjamin, the truth is that, while the practice probably doesn't have any deleterious effect on eyesight, definitive proof is lacking and, given the problems of performing such studies, likely to remain so.

? What were the health effects of the Chernobyl explosion?

It has been known since the 1920s, and the Nobel Prize-winning work of the biologist Hermann Muller, that radiation can cause severe genetic damage to living organisms. One might therefore expect that huge numbers of deformed babies were born in Ukrainian hospitals following the Chernobyl nuclear reactor disaster of April 1986. Surprisingly, there is no evidence for any genetic mutations among those born to parents exposed to the nuclear fall-out. To quote from the United Nations report on the long-term effects of Chernobyl, published in 2000: "No increase in birth defects, congenital malformations, stillbirths, or premature births could be linked to radiation exposures caused by the accident". Indeed, the UN expert committee found that while there were around 1,800 cases of thyroid cancer among children, "there is no evidence of a major public health impact attributable to radiation exposure fourteen years after the accident".

Presumably the level of radiation released, while far higher than natural background levels, was not sufficient to trigger inheritable genetic damage – or at least, not sufficient to stop the body's own DNA repair systems from repairing any damage.

One might think this was surely not the case in the atomic bombing of Japan in 1945. Since 1947, the Radiation Effects Research Foundation, based in Hiroshima, has monitored the health of tens of thousands of survivors of the bombing and the outcome of 70,000 pregnancies. In 1995, a search for genetic damage among 25,000 children involving many millions of genes found a total of just 6 mutations – and only two of those belonged to children whose parents had been close to the point of detonation. Since then, further studies have been carried out with the same result. In short, neither Chernobyl nor the atomic bombing of Japan has produced a single case of radiation-induced congenital birth defect.

? Is it true that aluminium pans can cause Alzheimer's Disease?

The first hints of a possible link between aluminium and Alzheimer's Disease emerged in 1965, following experiments in which rabbits injected with aluminium compounds developed tangled formations in their nerve cells. Concern heightened in the 1980s, when studies of the brains of elderly people who had died from Alzheimer's Disease detected the presence of aluminium. This led to fears that the aluminium had leached from cooking equipment and prompted many (including myself) to throw out their aluminium kitchenware. The current consensus is that the saucepan scare was unwarranted. The human body is very adept at ridding itself of the typical level of aluminium it ingests. Furthermore, studies of people exposed to relatively high levels of aluminium (through certain medicines) have failed to find an association with Alzheimer's. Still, the scare did at least provide an excuse for replacing grotty old pots and pans with some very nice stainless steel ones.

? What is the best way to burn off calories on the way to work?

It seems obvious that one should choose the most vigorous means of getting to work, which would point to running as the best way to burn off calories on the way there. Happily, for those of us who don't fancy arriving at work in Lycra shorts and a sweaty T-shirt, there is a get-out clause. Although running burns up more calories per minute than walking, it will also get you to work much more quickly and so you do it for much less time. The question then becomes whether the greater speed cancels out the higher rate of energy consumption – and it does. For example, a 60 kg person running at 6 mph burns up around 11 kilocalories per minute, which is around three times the rate used up by walking at 2 mph. However, running at that speed will also get you to your destination three times faster. So, if your place of work is a fixed distance away, the total number of calories burned off is hardly any different, whether you run or walk. If you're prepared to run flat out like an athlete, then your increased fitness may boost your "basal metabolic rate", the energy you burn up simply being alive, but the effect won't be dramatic. Add in the fact that walking puts a lot less strain on the joints, doesn't require fancy gear or a shower at the end and it's no contest – walk, don't run.

? What can you eat that uses up more calories than it gives?

The calorific value of a food is a measure of its energy content and at a fundamental level this depends on the relative numbers of carbon, hydrogen and oxygen atoms it contains. As a general rule, high calorie foods have relatively large amounts of carbon and hydrogen relative to oxygen, giving them more scope for combining with extra oxygen and releasing chemical energy. This is why fat, with its relatively high ratio of carbon and hydrogen to oxygen is much higher in calories than simple carbohydrates like sugar.

For a food to be a source of "negative calories", it must provide fewer calories than are used to consume it. Sadly, there are no zero calorie foods: all are combinations of oxygen, carbon and hydrogen, capable of releasing energy. So, in the search for a zero- or even negative-calorie food, a little more cunning is required. For example, one could consume foods whose energy content is never released, because the body cannot break them down. Cellulose is a case in point: rich in energy-containing carbohydrates in a form humans have lost the ability to digest, so it passes through with its calories untouched. This is why vegetables like celery and cabbage have such very low calorie counts of just ten to twenty kilocalories per 100 grams: they are packed with energy but it is largely in a form humans cannot break down and use.

These could become "negative calories" if the energy required to digest them exceeds the amount provided. It's often said that the act of chewing a stick of celery is enough to cancel out its energy content. In fact, the energy involved in mastication is tiny; what really matters is that used up in the act of digestion – the "thermic effect" of the food. Experiments suggest that digesting a vegetarian meal burns up around fifty to sixty kilocalories, so that a bowl of salad would indeed provide "negative calories"; supplying significantly less energy than is used to digest it.

Further negative calories could be obtained by washing the meal down with a chilled "diet" drink; the sweeteners used are pretty rich in energy but their extreme sweetness means that only tiny amounts are needed in a drink – which thus contains only a few calories per can. The negative calories then come from the fact that once we've consumed it, we expend about ten kilocalories warming the liquid up to body temperature. As the chilled drink contains around three kilocalories, drinking it will supply about seven "negative kilocalories".

Now the bad news: the amount of "negative calories" one could reasonably expect to consume in a day is probably no more than a few dozen, while it takes around 3,500 kilocalories to burn off a single pound of fat. In other words, negative calorie foods may exist but they're irrelevant in the fight against flab.

? Is it dangerous to drink out of lead crystal glasses?

Lead crystal is ordinary glass made more sparkling by the addition of twenty to thirty per cent lead oxide. The toxic properties of lead have prompted researchers to look into the potential health effects, with vaguely worrying results. In 1991, the *Lancet* carried a report by two pharmacologists from Columbia University, New York, showing that lead can leach out of crystal decanters, putting potentially harmful levels of lead into the wines and spirits they contain. I might add there have also been cases of lead poisoning among those who drink from pewter tankards, which contain around ten to twenty per cent lead.

? Is leprosy infectious?

Affecting chiefly the skin and peripheral nerves, leprosy is caused by the bacillus *Mycobacterium leprae* and can be caught from those carrying the disease, primarily by inhalation or physical contact. This is especially true in the case of lepromatous leprosy, which produces many skin lesions; tuberculoid leprosy, in contrast, is much less easily transmitted, as skin lesions are rarer. That said, it usually takes prolonged exposure to contract the disease and in any case most people have a natural resistance to the bacillus.

? Is it true that eating carrots is good for the eyes?

The supposed ability of carrots to improve eyesight generally and night vision in particular, came to prominence during the Second World War, in accounts of the uncanny ability of RAF pilots to shoot enemy planes out of the sky. There have been claims that the story was a Government-inspired plot to cover up the fact that British aircraft had been fitted with radar. If so, it may have aided the enemy more than its creators would have liked, as carrots do contain a compound that can help boost visual acuity.

The orange colour of carrots is a tell-tale of their relatively high content of beta-carotene, a hydrocarbon compound that is converted in the body into vitamin A, which is helpful in allowing the eyes to adapt to darkness and keeping the cells of the eyes healthy. Several studies have found that taking Vitamin A or beta-carotene supplements can combat night-blindness and cut the risk of macular degeneration, which affects visual acuity. Crucially, these studies typically involve doses far higher than you'd get in the standard diet: you would have to eat your way through a sizeable heap of carrots to get the same intake. So while it's true that carrots contain compounds that are good for vision, don't expect to have eyes like a hawk after eating them up.

? Is it possible to catch cancer from someone else?

The first hints that some forms of cancer are contagious emerged almost a century ago when experiments showed that chickens exposed to tissue from birds with a form of leukaemia also died of the disease. At the time, even the mere existence of viruses was bitterly disputed and it took decades for the idea of cancer-causing viruses to become widely accepted (Francis Peyton Rous, who pioneered the field, had to wait until 1966 to win the Nobel Prize, at the age of 87). Since then, around fifteen per cent of human cancers have been linked to viruses and thus identified as contagious; the true figure may be closer to thirty. Millions of people are now known to die each year after becoming infected with viruses; the Human Papilloma Virus (HPV) alone kills over 1,000 women in the UK each year, from cervical cancer. Doctors suspect that following the sexual revolution of the 1960s, many women became infected with HPV, leading to cervical cancer becoming the most common sexually transmitted disease among British women.

? Are copper or magnetic bracelets effective against disorders like arthritis?

Considering how long these bracelets have been about and how popular they are, one might expect to find a wealth of studies of their efficacy, but no. A trawl back to the mid-1960s turns up very little in the way of research and what has been done is suggestive but far from conclusive. An Australian study of several hundred arthritis sufferers, published in 1976, found evidence that many obtained genuine therapeutic value from wearing copper bracelets. Quite why is far from clear, however. Measurements suggested that trace levels of copper do migrate from the bracelet into the body but other studies have failed to show a clear role for copper in rheumatoid arthritis. Many doctors would no doubt dismiss any positive benefits as being the result of the placebo effect (which often produces improvements in as many as thirty per cent of patients for a wide range of disorders) but overall the issue of copper bracelets is an example of "more research needed". In the case of magnetic bracelets, a recent study by British doctors of around 200 men and women with osteoarthritis of the hip or knee pointed to some evidence of efficacy but again it's far from clear that it is more than a placebo effect (though of course that doesn't mean such bracelets are useless).

? Can you get sunburnt in a car with the windows up?

The cause of sunburn is often said to be the ultraviolet radiation in sunlight and this is true up to a point. The actual cause is a specific type of relatively short-wavelength ultra-violet radiation, UVB, which does not pass through glass. So, in strict scientific terms, you can't get sunburnt through car windows. That said, longer-wavelength, UVA, rays can still penetrate certain types of glass (including ordinary sunglasses) and while these don't cause the damage that leads to potentially fatal skin cancer, they can cause symptoms ranging from temporary reddening of the skin to premature ageing and wrinkles. In high-UV countries, dermatologists recommend fitting special UVA-absorbing films over all the windows, to minimize the risk of long-term skin damage.

Chapter 3

Beliefs, myths and mysteries

❓ Is it rational to believe in God?

Many great thinkers have addressed this question, most famously the philosopher Blaise Pascal, whose resolution of the question was first published, posthumously, in 1670. Pascal's Wager begins with four possibilities: God either exists or not and our belief in Him is either well-founded or not. Pascal attached "pay-offs" to the resulting combinations, arguing that it is rational to choose the combination offering the biggest pay-off. For example, if God does not exist, then believing in Him means wasting a lot of time listening to boring sermons; on the other hand, if He does, non-believers run the risk of incurring His almighty wrath.

The problem is that we don't know what our actual chances of getting these various pay-offs are, as we don't know whether God exists or not (if we did, we wouldn't be going through this rigmarole in the first place). Pascal dealt with this by introducing probabilities to the various options (in the process inventing a new branch of applied mathematics, known today as "Decision Theory"). His choice of probability was, however, very much open to question. Arguing that God is as likely to exist as not, Pascal set the odds on His reality at 50:50. Fortunately,

Pascal rendered his choice irrelevant by the further claim that belief in God combined with God actually existing produces an infinitely large pay-off – namely, spending eternity in Heaven. This trumps all other possibilities, regardless of their probability and allowed Pascal to claim it is therefore rational to believe in God.

Needless to say, many see this as mathematical sleight of hand and even three centuries later a vigorous debate continues about the relative merits of alternative arguments. For example, atheists have argued that the chance that God exists is precisely zero. In that case, Pascal's approach implies that belief is rational only if there's a bigger pay-off from wrongly believing God exists than from correctly believing He doesn't. Atheists might say God cannot possibly exist but legions of happy churchgoers insist otherwise. In any case, an insistence that the probability of God is precisely zero implies perfect knowledge, which many would argue is the preserve of, well, God.

Intriguingly, Pascal's approach shows that, under reasonable assumptions, belief in God is rational for all who believe they gain some benefit from such belief, *regardless* of whether it is well-founded or not.

Pascal's Wager is immensely rich in philosophical implications and, used judiciously, can be extended to cast light even on the rationality of, say, belief in transubstantiation or reincarnation. What this ingenious mode of argument does not – and indeed cannot – do is provide proof that any given belief is rational for everyone. This is because, contrary to what we are so often led to believe, such rationality is as much a matter of personal preference as cold logic.

? Is there any scientific evidence for reincarnation?

A key part of the belief systems of Hindus, Buddhists, Sikhs and others, around one in five people cleave to the notion of reincarnation. Sheer weight of numbers is hardly a reliable guide to truth but nor should it be lightly dismissed. Not that this has stopped most scientists – especially when confronted with yet another twenty-first century nobody claiming to have been Joan of Arc. An admirable exception is Professor Ian Stevenson of the University of Virginia, who has spent the last forty-odd years tracking down and analysing hundreds of cases of apparent reincarnation. Professor Stevenson has documented the claims as thoroughly as possible and made a special effort to rule out the standard explanation: that people making such claims have, wittingly or otherwise, acquired knowledge about their "past life" from books, documents and those who knew the deceased. This has led him to focus on people apparently bearing physical scars created during the violent death of their previous incarnation. These are anything from odd birthmarks to disfigured limbs on children consistent with the fate of their adult "incarnation". Summing up his personal experience on more than 500 cases, plus involvement in over 1,000 more, Professor Stevenson has concluded that reincarnation provides the best – though not the only – explanation. Sceptics have, quite reasonably, pointed out the possibility that the similarity of the scars between the deceased and their reincarnation could simply be coincidence, around which the people involved weave a *post hoc* account. Professor Stevenson counters this with the "crown jewels" of his research: a set of a few dozen cases of alleged reincarnation in which he has managed to interview those concerned, identify the deceased person and corroborate the supposed "memories" of the previous life. Predictably, sceptics say that he cannot prove the two families were never in contact, so these cases are hardly more impressive than the others. It will be clear from all this that reincarnation is an issue for which evidence is a necessary but not sufficient condition for acceptance by the scientific community. The problem is not a lack of evidence – Professor Stevenson has surely demonstrated that – but a dispute over its reliability, and this is a matter of trust, not science.

Faced with such situations, one can proceed in a number of directions. One can cut through the Gordian knot of complexity by rejecting the whole issue of reincarnation. Alternatively, one can turn to the philosophical device, *Occam's Razor*, named after

William of Occam, a fourteenth century philosopher-monk. This states that explanations making the fewest assumptions are also the most plausible (a claim that can be proved mathematically). Being essentially probabilistic, this rule will not always lead one to the right conclusion in any specific instance but will do so more often than not. Finally, one can show that (under reasonable assumptions), it is rational to believe in reincarnation if one is convinced that such a belief makes one a better and happier person here and now, *regardless* of whether the belief is well-founded or not. I am sure many Buddhists would go along with this and to that extent their belief is entirely rational.

？ Is it true water empties out of sinks anti-clockwise north of the Equator?

It's often said that water draining out of a sink or bath swirls down the plughole anti-clockwise in the Northern Hemisphere and in the opposite direction south of the Equator. The explanation is supposed to be the Coriolis Force, which affects anything that tries to move across the face of a spinning object – such as bathwater on the rotating Earth. The Coriolis Force plays a key role in making atmospheric depressions and hurricanes coil in an anti-clockwise direction as they move north of the Equator. The trouble is that it is very feeble and the resulting acceleration only produces significant effects over relatively long time-scales. In the case of water draining out of a bath, the Coriolis Force produces an acceleration ten million times weaker than gravity and its effects are thus easily masked by, for example, a very slight tilt of the bath.

That said, in 1961 the science journal *Nature* reported that Dr Ascher Shapiro, in Massachusetts, had successfully detected the Coriolis effect in extremely careful experiments, which showed water flowed anticlockwise out of a two metre wide tank. In 1965, a team at the University of Sydney, led by Lloyd Trefethen, repeated the experiment south of the Equator and found that the water did indeed swirl out in the other direction.

Surprisingly, the Coriolis Force also has a significant effect on anyone taking a stroll – or at least it would, were it not for the friction between our feet and the road. A rough calculation shows that someone walking for a mile down a road would be driven off-course by around 180 yards. Thus, if the road were icy and we slid down it, we would find ourselves having to take evasive action to counter the effects of the Earth's rotation (there are claims – probably apocryphal – that Antarctic penguins tend to walk in paths that curl round to the left).

? Can opera singers really shatter wine glasses by singing?

The renowned soprano Dame Nellie Melba is reputed to have performed this trick and in 1971 the audio-tape maker Memorex based a very successful advertising campaign around Ella Fitzgerald shattering glass both when singing live and when recorded on its tape cassettes (a claim the company still stands by today). Even so, it is hard to believe that mere vibrating air can shatter solid glass. The explanation lies in the simple yet extraordinarily powerful phenomenon of resonance, according to which even relatively weak vibrations can perform miracles, if tuned to the right frequency. A classic example is the playground swing: even a small child can propel an adult many feet up in the air by pushing at the right moment. In the case of a wine glass, the trick lies in using fine crystal glass, (which has a relatively rigid molecular structure) tapping the glass to hear its natural ringing frequency and then hitting that note and sustaining it. If a singer's voice is both stable and strong enough, it will set up resonant sound waves inside the rim of the glass that eventually shatter it.

All of which prompts one to wonder whether something similar might explain the Old Testament story about the walls of Jericho being brought down by the sound of horns and human voices. It seems unlikely: neither the nature nor the shape of walls are suitable for setting up the all-important resonance. Sound consultant David Lubman of Westminster,

California recently examined the possibility that the walls were flattened simply through raw noise. He calculated that to damage Jericho's stone or mud-brick walls the acoustical power required would be almost a million times greater than even a generous estimate of the power produced by the horns and besieging army. This led him to suspect that the legend is a metaphorical reference to the psychological effect of the sound, which might have emboldened the army while putting the wind up the city's defenders.

? Are old windows thicker at the base because glass flows like syrup?

Glass is often referred to as a "solid liquid", the molecules of which are held together relatively loosely, like extremely thick clear syrup so it seems reasonable to think old glass would sag under the effect of gravity, making panes thicker at the bottom. The trouble is, glass isn't really very runny at all: according to calculations published in the *Proceedings of the Royal Society* in 1999 by Dr Yvonne Stokes of the University of Adelaide, it would take at least ten million years for a window pane to become just five per cent thicker at its base. Yet panes of ancient glass do seem to be thicker at their base. One plausible explanation is that medieval glaziers exploited the fact that glass makers could not make truly flat panes of glass by standing panes up on their thicker ends when installing them.

? Is there any truth in reports of the Indian Rope Trick?

Oddly enough, while everyone seems to agree that the Indian Rope Trick is precisely that – a trick – no conjurer has ever figured out how it works. In its standard form, it involves a fakir persuading a rope to stand upright with no visible means of support. A young boy then clambers up to the top of the rope and vanishes. After failing to persuade the boy to return, the fakir then climbs the rope himself and also vanishes. Legends telling of similar miracles – such as twine ascending into the air – have been circulating in the East for centuries, but the trick itself only became famous towards the end of the nineteenth century, when an eyewitness report appeared in the *Chicago Daily Tribune*. On 9 August 1890, Fred S. Ellmore described how a juggler in India had caused a rope to rise into the air from the ground, up which a boy climbed and duly vanished. According to the reporter, the trick bore witness to the powers of mass hypnosis.

Four months later, the newspaper itself confessed that the entire story was a hoax designed to boost circulation (take another look at the reporter's name). It made no difference: by that time, both the "eyewitness account" and the "explanation" had gone round the world, prompting yet more sightings and in 1919, photographic evidence.

In their study of the history of the Indian Rope Trick, published in the *Journal of the Society for Psychical Research* in 2001, Dr Peter Lamont of the University of Edinburgh and Dr Richard Wiseman of the University of Hertfordshire concluded that reports of the Indian Rope Trick are not entirely fraudulent; rather, they appear to be embellishments of a much simpler trick, involving a pole with a child balancing at one end.

Even so, despite modern magicians being able to make the Statue of Liberty disappear or mountains move, no one has found a way of performing the Indian Rope Trick in its original form. While many magicians have succeeded in persuading ropes to rise and children to vanish on stage, no one has managed it in the open air, on the ground, in broad daylight. Perhaps the illusionist David "man in a plastic box" Blaine could be persuaded to try his hand.

? Has anyone succeeded in creating a new form of life?

This is one of those scientific breakthroughs people have been talking about since the night Mary Shelley ate cheese too close to bedtime and dreamt up Frankenstein's monster. Oddly, the fact that it has been achieved seems to have passed most people by – perhaps because it turned out to be a bit of an anti-climax. In November 2002, a team led by Professor Michael Travisano at the University of Houston reported the creation of an entirely new species of life in the laboratory from the combin-ation of two types of yeast, the single-celled organism famed for turning sugar into alcohol. Unlike mere hybrids, the resulting lifeform can reproduce but, like a true species, it can only breed with others of its own kind. Creating the new species took just four weeks, prompting suspicions that nature may use a similar method to create new species outside the laboratory.

? How long is "once in a blue moon"?

People have been using this phrase to imply "hardly ever" for at least 400 years. It seems to have its origins in a rare meteorological phenomenon in which the moon's light passes through dust or smoke particles, creating a bluish tinge similar to that caused by cigarette smoke. In the last 150 years or so the phrase has somehow acquired a second meaning: a month in which two full moons appear – a calendrical quirk made pos-sible by the fact that the average length of a month is just a smidgen longer than the period between successive full moons. Quite why this has anything to do with the colour blue isn't clear but it does at least lead to a precise definition of the phrase: a blue moon puts in an appearance once in a period of slightly over two years.

? What are the origins of the phrase "as mad as a hatter"?

In his wonderfully entertaining *Nature's Building Blocks* (OUP, 2001), John Emsley ascribes the phrase to the effects of chronic exposure to mercury vapour. According to Dr Emsley, the animal pelts used in hat making were dipped in mercury nitrate and then dried, to bring the fur to the right consistency.

At a time when mercury was widely prescribed for everything from skin complaints to syphilis, no one worried much about the inhalation of the fumes. Mercury has since revealed itself to be a powerful neurotoxin and in retrospect the effects of years of exposure to its fumes were plain for all to see and manifested themselves as anything from the "hatter's shakes" and insomnia to depression and full-blown paranoia.

❓ Is there a truth drug that is 100 per cent effective?

The idea that certain potions can perform miracles such as increasing lifespan, triggering desire or revealing truth refuses to go away, despite all evidence to the contrary. There have been claims that US intelligence agents recently used one such "truth drug", sodium pentothal, to extract information from Abu Zubaydah, a senior al-Qaeda activist. This may be true, as might reports that he began giving details about the structure of the terrorist network. What is rather more dubious is the reliability of any information thus gleaned.

The notion of a "truth drug" first emerged in 1931, when the *American Journal of Police Science* carried a report about the effects of scopolamine, a potentially lethal extract of deadly nightshade. The report claimed that, when given to women to ease the pain of childbirth, the drug made them much more talkative and uninhibited. This led to an experiment in which scopolamine was given to two prisoners before their trial. Both protested their innocence and were subsequently acquitted. In a stunning display of faulty logic, this was taken to be evidence that the drug had revealed the truth – and the myth was born.

Concern over the side-effects of scopolamine led to barbiturates being used to mimic the depressive effect thought to trigger confessions and Brevital, sodium amytal and especially sodium pentothal have since been touted as truth drugs. Reviewing the scientific evidence in his classic text *The Psychology of Interrogations, Confessions and Testimony* (Wiley, 1992), Dr Gisli

Gudjonsson of the Institute of Psychiatry, London, concluded that while such drugs may increase the talkativeness of those who are already chatty, they have little or no effect on those with no intention of revealing anything. He added: "There are serious problems with the reliability of information obtained during the procedure due to increased confabulation, fantasy and suggestibility among vulnerable subjects". Still, why let such scientific hand-wringing get in the way of an intelligence coup?

? Is there any scientific basis to the idea of "brainwashing"?

The concept of brainwashing seems to have largely been an attempt by Cold War-era Americans to understand how otherwise normal people could possibly believe in Communism (the term first appeared in a Florida newspaper in September 1950). The thesis was that the people of Russia, China and other Eastern bloc countries must have been stripped of their personalities and "programmed" into blind obedience towards the state.

Paranoia about the power of brainwashing surged at the end of the Korean War, when some American POWs came home denouncing aggression and praising Communism. Within the CIA, however, most concern centred on the possibility that America had been left behind in deploying a new form of warfare.

As Robin Waterfield explains in his fascinating review of the evidence in *Hidden Depths: the story of hypnosis* (Macmillan, 2002), the CIA spent years trying to develop brainwashing methods but only succeeded in showing how ineffective they were. Initial experiments centred on hypnotising people into performing illegal operations, including murder. Secretaries were persuade to steal documents and fire unloaded guns at people but the researchers concluded this said more about the compliance of secretaries than the effectiveness of hypnotic brainwashing. Attempts to persuade people to perform such acts after hearing or seeing a "trigger" signal (as in the 1962 film *The Manchurian Candidate*) gave hopelessly unreliable results.

Most effort was spent on brainwashing via more forceful means. Denial of sleep, electric shocks, sensory deprivation and psychoactive drugs were all used, often illegally: in 1975, it emerged that the CIA had used LSD on over 1,000 people without their consent. Even so, the results were too unpredictable to be of any value: some victims simply became insane and committed suicide.

If there is any evidence for the effectiveness of brainwashing, it comes from the CIA's own actions, the organization apparently having brainwashed itself into justifying such egregious experiments.

? Does subliminal advertising work?

There is something rather disturbing about the idea that our actions could be influenced by ideas slipped into our unconscious mind. Not surprisingly, there was uproar in 1957 when the New York-based Subliminal Projection Company claimed to be able to boost sales by slipping subliminal messages into adverts. At a press conference, the company unveiled the results of an experiment set up in a local cinema, where the slogans "Drink Coke" and "Eat Popcorn" had been slipped into a film, appearing for less than a thousandth of a second. According to the company, while no one noticed the insertions, Coke sales went up eighteen per cent and popcorn sales zoomed by well over fifty per cent. In the resulting storm of protest, psychologists condemned the experiment as "nightmarish" and broadcasting regulators moved to ban all subliminal methods. In the UK, such adverts are still specifically forbidden under the terms of the 1990 British Broadcasting Act. However, the subject still rears its head from time to time. In the 2000 US presidential campaign, Republicans were accused of slipping the word "rats" into a television advert featuring images of Al Gore (they denied the charge, and claimed it was simply part of "bureaucrats").

Despite the outrage engendered by these supposed assaults on our free will, there is no evidence suggesting subliminal adverts are any more effective than conventional ones. While the brain can certainly detect messages that are too fleeting for conscious recognition, any response to them still has to enter the conscious mind, which retains the ultimate power of veto.

? How did tribes make "shrunken heads"?

Of the many bizarre items on display in the wonderfully eccentric Pitt Rivers Museum in Oxford, nothing tops the ten shrunken heads or "tsantsas" (they're in the case labelled "Treatment of Dead Enemies" – you can't miss them). They were made by various tribes from the Upper Amazon region, in the belief that taking and processing the heads of enemies pacified their spirit and formed a bond with the tribe which took them.

Despite appearances, shrunken heads aren't really proper heads but merely the outer skin; the skull and its contents are removed before a brief boiling and the skin is then dried by stuffing it with hot pebbles and sand. After darkening the face with vegetable dye and sewing the eyes and mouth shut, the shrunken head is ready for display. According to the Pitt Rivers Museum, the practice of making shrunken heads continued in some parts of the Amazon until the 1960s.

? Is it true that it is unwise to use the Internet during an electric storm?

In general, all electrical appliances, including computers, should be unplugged when a violent storm approaches (but perhaps left alone once it's overhead, lest lightning strikes while one is touching the equipment). Computers connected to the

Internet are, however, particularly vulnerable and should be disconnected from the phone line, as this is the route by which the most damaging surges pass into the machine.

❓ Is it true that the right foot is more ticklish than the left?

During the early 1980s, for reasons best known to themselves, a team of Italian scientists carried out a series of studies of the ticklishness of people under a variety of circumstances. It was during one of these, involving forty-six female psychology students, that researchers found that people seem to have one foot that is more ticklish than the other – and in most people it is the right.

The experiment was repeated in 1998 by two researchers at Stirling University, using a special device to ensure a consistent stimulus: a pointed nylon rod was stroked across the sole of the foot three times at second intervals. After applying it to thirty-four people, the researchers confirmed the original finding and pushed back the frontiers of knowledge still further by showing that males were more ticklish than females.

One suggested explanation is that the left side of the brain – which detects stimuli applied to the right foot – is associated with positive emotions, such as laughter. As the difference was not very large, my suspicion is that this is a bit of *post hoc* rationalization of an effect that doesn't really exist. Even so, it seems likely that everyone had a lot of fun doing the research.

❓ Have scientists found the "missing link" between us and chimps?

No – not least because there isn't one. Fossil and DNA evidence has shown that, contrary to the caricature of Darwin's theory, humans are not descended from chimpanzees. Rather, we shared a common ancestor until, around five to seven million years ago, humans and great apes began to evolve in different ways. Quite what this common ancestor looked like is highly contentious.

? What is the current best theory to explain the Star of Bethlehem?

To my mind, the most convincing explanation is the one that does the best job of explaining the Star's sudden appearance, location and brightness. In 1954, Arthur C Clarke suggested that the event may have been due to a supernova, the detonation of a super-massive star. A report of just such an event turned up in 1977 in records of celestial events kept by Chinese astronomers, which suggest a bright star suddenly appeared near the constellation Capricorn in 5 BCE and lingered in the sky for seventy days before fading. This ties in well with the timing of the birth of Christ as deduced by historians; the problem lies in that figure of seventy days. Astrophysical theory provides a link between how bright a supernova is and how long it takes to fade and for the "Christmas supernova" to have faded so quickly, the original explosion would perforce have been a bit of a damp squib.

What seems to me the best theory yet was put forward in 1999 by Mark Kidger, who argued for the Star being a nova, a kind of stellar explosion triggered when hydrogen is dragged from one star on to another orbiting around it. The nova theory fits in well with both the Gospel accounts and known physics and is also backed by Chinese astronomical records, which seem to point to a nova explosion in the constellation Aquila in mid-March, 5 BCE.

? Is there any evidence for a special "link" between identical twins?

There is certainly a lot of anecdotal evidence pointing to some form of psychic link between twins, both identical and fraternal. When Ross McWhirter was shot dead by IRA gunmen one November evening in 1975, his identical twin Norris collapsed at his home thirty miles away; his family thought he had suffered a heart attack. As parapsychologist Guy Lyon Playfair recounts in his book *Twins: the psychic connection-telepathy's special case* (Vega, 2002), recent studies suggest that telepathic experiences are three times more common among twins than among non-twins and that many twins report experiences such as suffering "sympathetic" pain.

Attempts to probe the supposed links have met with only mixed success. In 1993, a study by parapsychologist Dr Susan Blackmore found that twins asked to draw whatever came into their minds were more likely to produce similar drawings than non-twins but she concluded that this was because twins tend to think alike, rather than share their thoughts telepathically.

Other tests have looked at whether changes in the brain activity of one twin can trigger a similar change in their sibling. In 1965, the prestigious US journal *Science* published the results of a CIA-backed experiment suggesting the brain wave patterns of identical twins could show signs of such linkage; the evidence was hardly compelling, however. More impressive results emerged in 1994 from a team at the National Autonomous University, Mexico, which found evidence of linked brain activity in around twenty-five per cent of people with a close emotional bond.

In short, there is some evidence suggesting that twins – especially emotionally close ones – may have some special "link" but in the absence of a large, thorough study, it's far from compelling.

? What is the evidence supporting claims of "spontaneous combustion"?

In the 150 years since Dickens described the horrific death of Krook the rag-dealer in *Bleak House*, several hundred cases of apparent spontaneous combustion in humans have been recorded. They typically involve the rapid yet complete incineration of the person, often with no obvious nearby source of heat.

The scientific response (other than the standard one of dismissing it all as the ravings of morons) has centred on the "wick effect" in which the victim's clothes catch alight on a candle or dropped cigarette and act like a candle wick, heating the body to temperatures where body fat melts, giving the fire a source of fuel for hours. This was tested in 1999 by John de Haan, an American arson investigator, who used a pig carcass to show that the "wick effect" can generate temperatures in excess of 760°C. While this is certainly enough to explain the astonishing thoroughness with which bodies are burned during spontaneous combustion, it does nothing to explain its rapidity: there are documented cases of people being turned into ash in a matter of minutes. In short, scientists are now willing to concede that spontaneous human combustion is a genuine phenomenon but still have some way to go before explaining all its paradoxical properties.

? Is it true that we lose twenty per cent of our body heat through our heads?

The head makes up less than ten per cent of the total surface area of the body but its high density of blood vessels means that it's responsible for a much higher proportion of the total heat loss. Exactly how much is hard even to guesstimate and I am indebted to Dr Dusan Fiala of De Montfort University, Leicester, who investigated the question using his computer model of heat loss from the human body. Putting in the figures for a fully-clothed person taking a walk on a cold winter's day (0°C), Dr Fiala reported that the body loses heat at a rate of around 260 watts, of which about sixteen per cent comes directly from the surface of the head. Add in another few per cent from the effect of breathing out warm air and the figure of twenty per cent is certainly of the right order of magnitude. To find out the effect of wearing a hat, I asked Dr Fiala to put a virtual trilby on the head of his computerized person. The effect was dramatic, with the heat loss directly from the head falling by fifty per cent – amply confirming the wisdom of wearing a hat in cold weather.

? Is it true that string vests are better than ordinary vests?

In theory at least, string vests are supposed to be warmer because they act as a cotton "frame", forming pockets of insulating air between the skin and whatever is worn over the vest. I've never heard of any scientific evidence that this is actually true in practice (come to that, I've never seen a string vest on sale either). I suspect they're not very effective, as the air pocket effect would constantly be undermined when the clothing lying over the vest breaks contact with it. On the other hand, my youngest son Theo is happy to confirm that wearing many thin layers – such as fifteen T-shirts – works extremely well, even if he does end up looking like a tiny Michelin Man.

? Is it possible for your brain to hurt if you think too much?

Curiously, the brain lacks pain receptors and so can't experience pain itself, which allows neurosurgeons to study brain function using fully conscious patients, using just a local anaesthetic when opening up the skull and tissue around the brain. So while our legs or arms give us a lot of pain after a hard workout, our brains can spend hours pondering the intricacies of a tax return without discomfort. The real source of "brain-ache" is most likely to be prolonged tension in the neck and scalp muscles caused by spending too long in a tense, hunched position.

❓ Is it true that Mt. Everest is not the highest point on Earth?

It all depends what you mean by "highest". The usual definition is height above mean sea-level, in which case there is no doubt that Mt. Everest is, at 29,035 ft, the highest point on the planet. Some have, however, argued (chiefly out of contrariness, as far as I can tell) that the highest point should be considered to be that furthest from the centre of the Earth. In that case, one must take account of the fact that the Earth's rotation causes a bulge at the Equator of around 14 miles relative to the poles.

As it happens, Ecuador's Mt. Chimborazo lies within 1° of the Equator and though its summit is only 20,700 ft above sea-level, it benefits mightily from the bulge effect compared to Mt. Everest and its summit stands over 7,000 ft further from the centre of the Earth.

Regardless of all this geodetic wrangling, as far as I am aware, Mt. Everest remains unchallenged as the best vantage point on the planet. The rule of thumb for calculating the distance to the horizon in miles is to take the altitude in feet, add on fifty per cent, and take the square root. As Mt. Everest stands only 13,000 ft above the immediately surrounding land, that could affect the view from the top. However, the land to the south-east falls away fast enough to guarantee an uninterrupted view of around 210 miles from the peak. The effects of atmospheric refraction may even increase this to 230 miles or more.

❓ Why does putting a spoon into a bottle of sparkling wine help retain the fizz?

Some people insist that a silver spoon works best. I've tried this myself and it's absolutely true: a bottle of sparkling wine with a spoon in its neck retains its fizz remarkably well. The thing is, so does a bottle without a spoon in its neck. In other words, if not too shaken up, sparkling wine keeps its fizz perfectly well for a day or so with no outside help at all.

? Is it true that we typically use only ten per cent of our brain?

The brain contains around 100 billion neurons, each with about a thousand connections and doubtless most of us fail to make full use of this prodigious computational and memory capacity. However, while some parts of the brain are more active than others when performing specific tasks, brain scans show that the whole brain is active, at some level, most of the time.

? Could you travel round the world by ascending in a helicopter and waiting for your destination to arrive beneath you?

The surface of the Earth rotates at around 1,000 mph at the Equator, twice as fast as the average 747. So how come no airline is offering cheap, no-frills "hovers" to the destination of your choice? The reason is that even a helicopter that seems to be sitting on the ground is sharing the rotation of the rest of the planet and when it takes off in a vertical hover, it carries that forward speed with it, thus keeping pace with the ground below. It is possible to do better by floating up high enough to be picked up by the high-altitude air currents generatedby atmospheric pressure. These blow at up to 400 mph, relative to the surface of the Earth; in 2002 they allowed balloonist Steve Fossett to zoom right around the world, with no visible means of support or propulsion, in just thirteen days.

? Where do all the odd socks go?

A few years ago, Professor Stephen Hawking gave a public lecture in which he declared his belief that socks disappear down mini black holes that appear spontaneously through quantum gravitational effects. This may sound somewhat desperate but it does have the advantage of accounting for the random disappearance of ballpoint pens, combs and children's shoes. It would also explain why collecting odd socks in a bag in the hope of reuniting them with their partners makes no difference: they never turn up again, as they have left our universe entirely.

More seriously, it is possible to do some mathematics on the problem of odd socks and the results do seem to explain the astonishing speed with which they accumulate. Imagine one sock goes missing at random – down a black hole, the back of the sofa or whatever – thus leaving an odd sock in the drawer. Now, imagine another sock goes missing. This could be either the odd sock already created or – far more likely – a sock from an as yet unbroken pair. It is thus clear that when a second sock goes missing, this is most likely to lead to the creation of yet another odd sock in the drawer, leading to a build up of odd socks. One can derive a formula that captures this effect; the results are salutary. For example, if we began with fifteen pairs of socks, by the time we have lost half of them at random, the most likely result is that we will be left with just three complete pairs lost among nine odd socks. No wonder they're so hard to find in the morning. The solution is to either buy socks of identical design or ensure that the socks stay together at every stage from foot to laundry basket to drawer. As many have discovered, the "variety packs" of three different designs of sock one gets from unimaginative relatives at Christmas are especially prone to this effect. Indeed, one can prove that the socks making up such packs always disappear in such a way as to maximize the number of odd socks left behind.

❓ Why is spinach so good for us?

While it's a perfectly good vegetable, the status of spinach as a wonder food has its origins in a blunder made by a German nutritionist in 1870, who, while recording its iron content, put the decimal point in the wrong place, boosting the value by ten times. The mistake wasn't discovered until the 1930s, by which time sales in the US had zoomed over thirty per cent thanks to the spinach-propelled exploits of Popeye.

? Where is the exact centre of Britain?

It all depends on what you mean by "centre". And as the Ordnance Survey told me while I was researching the answer, it also depends on what you mean by "Britain". According to Kevin McLoughlin of the OS, cartographers usually define the "centre" of a country to be its centre of gravity – in simple terms, its balance-point. So, one can make a rough estimate of the centre of Britain by drawing its outline on a piece of card, cutting it out and suspending it, one at a time, from a series of points around its edge. Lines drawn vertically downwards from each suspension point intersect at the centre of gravity.

The precise location of the centre depends on whether our retinue of islands is taken into account. If it is, then, according to the OS, the centre of Great Britain lies a mile to the south-east of Walker Fold in Lancashire. If the islands are left out, the centre of Great Britain moves to a point around three miles south-west of Clitheroe.

In the case of England alone, I long believed the centre to be in the village of Meriden near Coventry, which has a 500-year old monument proclaiming its status. Yet, according to the OS, the rightful claimant is the village of Fenny Drayton in Leicestershire (for real sticklers, the actual point lies a mile to the east of the village). Mathematical definitions aside, many people regard the centre of England as the point furthest from the sea. According to the OS, this is a point around a mile south-east of Coton in the Elms, a village between Swadlincote and Lichfield, around 65 miles from the Wash.

❓ What is the reason behind using vinegar to treat a wasp sting?

As the old rhyme goes, "Vinegar for a Vasp sting, bicarb for a bee", the idea being that vinegar, a very weak acid, will quickly neutralize an alkaline compound. Which is all very well, except that a wasp sting isn't alkaline, so putting vinegar on it is pretty useless. Some medical authorities say that a paste of bicarbonate of soda and water works for both bee and wasp stings; readers may like to investigate its efficacy. It is far more important, however, to extract a bee sting rapidly, without squeezing the poison sac, wash the affected area and watch for swelling or breathing difficulties or other signs of a potentially lethal allergic response. These affect around one per cent of children but are twice as common in the over-40s.

❓ Is it true that the world's population could all be crammed on to the Isle of Wight?

According to United Nations estimates, the population of the world is currently around 6,200 million. Let us give everybody enough room to stop them complaining they don't have enough room to swing a cat (which experiments with my neighbour's cat suggest requires a circular space about 11 ft across). Multiplying up, that means that the world's population could comfortably fit in a space of around 21,000 square miles; so Sri Lanka, Tasmania or Haiti would all fit the bill. If you insist on cramming everyone together to the point of standing room only, then it's possible to pack everyone in the world into a space of just 800 square miles, which is over five times the total area of the Isle of Wight. So no, you couldn't do it.

? Is it true that there are more people alive in the world today than have ever lived?

The idea that there are more people alive today than in all history has seeped into popular culture. It's not true, though – not by a long chalk. The population of the Earth today is around 6,200 million and it is true that this is the highest it has ever been: double the population of 1960 and six times the 1800 figure. The thing is, while the global population has exploded in recent decades, it has been growing at a pretty prodigious rate for several centuries and when totted up this represents an awful lot of individual lives. To estimate how many, demographers have used historical data to come up with graphs showing the population of the Earth at different times. These typically show a line with a gentle but steadily rising slope up until the mid-eighteenth century, after which the figures start to soar. Turning this into an estimate of the total number of humans who have lived calls for a few assumptions to be made, including of course a date for when the human race began. For example, the distinguished Canadian demographer Professor Nathan Keyfitz has opted for a date of 1 million years BCE as the time when there were just two humans on the planet. While many would no doubt quibble, it hardly matters: the final figure for the total number of humans is dominated by the huge population size and growth-rate over the last few centuries, which is known with reasonable accuracy. Cranking through the maths, the upshot is an estimated figure, for the total number of humans who have ever walked on the Earth, of around 60 to 100 billion. In other words, the people alive today represent no more than around ten per cent of all the humans who have ever lived.

By a curious coincidence (first noted, I think, by the science fiction writer Arthur C Clarke), there are also around 100 billion stars in our galaxy. So for every human who has ever lived, a star shines.

? Is it true that the best way to avoid being struck by lightning is to lie flat on the ground?

This is excellent advice, based on sound physics – but only as far as it goes. Anyone who stuck rigidly to it could find themselves being made rigid on a permanent basis. During a thunderstorm tall, pointy objects, such as soldiers tramping across relatively flat terrain, are in danger of acting as human lightning conductors. The electric field created by a storm tends to concentrate around point-like objects, which increases the risk of the insulating properties of the surrounding air breaking down, allowing electrons to flow between the object and the overhead cloud and triggering a lightning strike. By lying flat on the ground, the electric field around the body becomes relatively spread out, reducing the risk of being struck but it's still possible and then could prove fatal. The best advice is to run for shelter – a car is ideal – avoid trees with their myriad pointy bits and keep hunched up and low. If your hair starts to tingle and stand on end, it means you're about to join the ranks of those who have been struck by lightning. In that case, stop and put your hands on your knees: the shock should then pass through your arms and into the ground, without passing through your heart. With luck, you should then avoid becoming one of the ten per cent or so who fail to survive the experience.

Chapter 4

Numbers, games and pastimes

❓ Is there a good way to remember PINs?

Security experts are quick to tell us not to write down our PINs but have signally failed to come up with any practical alternative. Still, if they were so smart they would have recognized that a four-*letter* code produces far more combinations and is also easier to memorize.

Some people simply change the number to something more memorable, like a birthday. This isn't recommended, as dates do not permit certain number combinations, making it somewhat easier for a fraudster to find the correct digits. Similar caveats apply to adding a "memorable" number to a PIN and only writing down the result: the memorable number may well be more easily guessed than you might think. As a graduate student, the Nobel Prize-winning physicist Richard Feynman broke into safes containing the plans for the atomic bomb by guessing that the combination was 2718 – the first digits of a number widely used in theoretical physics.

Then there are the various ways of disguising PINs, such as keeping it as a fake phone number in a mobile phone or address book (which sounds a little risky, as fraudsters may not be averse to ringing them all up to find the fake). Many methods assume words are easier to remember than numbers and provide a way of converting the words to the PIN. My preferred method is secure, effective and requires nothing to be written down. Simply memorize a phrase constructed from words whose length corresponds to the PIN. Thus, if your number is 4315, the phrase could be "PINs Are A Chore". Zeroes can be dealt with using synonyms such as "nothing", "none" or "love". So if your PIN is 3036, you could use the phrase "It's nothing too clever".

? Is there a simple way of estimating the value of a pound in the past?

This amounts to requiring a simple estimate of inflation rates at different times. The trouble is that the definition of inflation, as well as its rate, has changed even in recent times. During the mid-1970s, the UK's inflation rate reached twenty-five per cent per year, around six times the rate at the time of writing (which is the lowest since the early 1960s). Thus, there is no simple formula into which one can plug a year and extract the appropriate inflation rate. Economic historians have prepared various estimates, dating back to the sixteenth century, and these allow some values to be extracted. So, roughly speaking, one pound today was worth twice as much in 1983, five times more in 1975, ten times more in 1968, twenty times in 1949, 100 times in 1727 and 500 times more in 1500.

? Why does multiplying two negative numbers give a positive result?

It is pretty clear why adding one negative number to another should produce an even bigger negative number (ask anyone with a overdrawn bank account). It also seems reasonable that negative numbers multiplied by positive ones should produce negative results (double an overdraft is an even bigger overdraft). Yet why the sudden change of sign when multiplying two negatives? This is perhaps most easily explained by picturing positive and negative numbers lying along a line: draw a horizontal line across a piece of paper and mark its mid-point; this is the location of zero. At equal intervals to the right of zero, write the numbers 1, 2, 3 up to 10; these are the positive numbers. Now do the same to the left of zero, –1, then –2, –3 and so on. Multiplication then becomes the process of moving in a particular direction from zero in a given number of jumps, each of a certain size. For example, to multiply 2 by 3, start at zero and face towards the right – as 2 is a positive number – and move by 2 sets of 3 units in the same direction (as 3 is also positive). You end up at 6 to the right – positive – side of zero. On the other hand, to multiply –2 by 3, start at zero but this time face left, as the 2 is negative. Then move by 2 sets of 3 units in the same direction, as the 3 is positive. Following the same conventions, to multiply –2 by –3, face left (because the 2 is negative), but move from zero in 2 sets of 3 in the *reverse* direction, because the 3 is also negative. This puts you at 6 on the right-hand (positive) side of the zero. There are of course more formal demonstrations and some involving bank debts but I find these contrived, as well as depressing.

? Why do things often split 80/20 – for example, 80 per cent of sales from 20 per cent of products?

This odd little ratio pops up, with suspicious regularity, in everything from charity campaigns to management training manuals. For example, according to the organizers of "Buy Nothing Day", developed nations, making up twenty per cent of the world population consume over eighty per cent of the Earth's natural resources, while around twenty per cent of insurance claims are said to account for about eighty per cent of the total amount paid out.

With no obvious reason why causes and effects should divide in this way, it is tempting to dismiss the 80/20 rule as a headline-grabbing device. Yet its ubiquity has intrigued economists for over a century and glimmerings of an explanation are starting to emerge. Attention has focused on a special case, known as Pareto's Law, after the economist Vilfredo Pareto, who in 1895 claimed to have found a mathematical formula governing the distribution of wealth. Known technically as an "inverse power-law relationship", the formula involves a quantity raised to a power which typically lies between 1 and 2.

Using economic data for his own country, Pareto found that the formula implied that twenty per cent of Italians owned eighty per cent of the country's total wealth. Astonishingly, he also found that the same split applied to economies as diverse as those of industrialized Britain and agrarian Russia and across the centuries from Victorian societies back to the days of the Roman Empire.

Not surprisingly, Pareto's claim to have found a "universal law" in economics was widely derided at the time but now it seems he was on to something. Recent research has shown that Pareto-like laws emerge naturally from economies that obey some pretty innocuous demands – notably that wealth is the result of exchange of money and goods between those taking part in the economy, plus random gains and losses due to speculation, property investment and the like.

It turns out that the ebb and flow of wealth through such economies leads to just the kind of "power law" found by Pareto. There is an interesting twist, however: contrary to his belief, the value of the power need not be fixed between 1 and 2. That, in turn, means economies need not always feature iniquitous 80/20 splits in wealth distribution. Studies suggest a fairer distribution of wealth can be generated simply by encouraging trade across the whole spectrum of incomes, from the poorest to the richest. All of which makes "Buy Nothing Day" appear even dafter than it sounds.

❓ Is it true that you can't prove a negative?

This is often wheeled out by people on the losing end of an argument but it is not true – at least, not in matters of logic. Indeed, the proof of negative statements is often used, in both logic and mathematics, to make major advances. The technique is called proof by contradiction and works by assuming the truth of the opposite of the original proposition and then proving that this leads to a contradiction. One of the finest examples of the power of this technique dates back to Euclid's proof by contradiction that there is an infinite number of primes. He began by assuming the opposite – that prime numbers stop at some maximum value – and showed that this implies there must be a still-bigger prime. This is a contradiction of the original assumption and thus proves that the primes go on forever.

Outside the formal world of logic and mathematics, the situation is less clear but more interesting. Among scientists, it is almost an article of faith that one can never do anything else *apart* from prove a negative – that is, prove that a theory is incorrect – as a watertight demonstration of its validity in all places at all times is patently impossible. This was first pointed out by the eighteenth century philosopher David Hume but became best known through the works of the twentieth century philosopher of science, Karl Popper. His arguments are often portrayed as confirming the belief that scientific theories can only be disproved but this is not true: scientists routinely dismiss challenges to their pet theories simply by appealing to other explanations, from questionable experimental technique to outright fraud. Scientists usually abandon their ideas not because they have been definitively proved incorrect but because of the arrival of another theory with more explanatory power.

? What's the biggest possible number?

The simple answer is infinity but in the 1870s, the mathematician Georg Cantor produced ingenious arguments revealing the existence of lots of different infinities – some of which are much bigger than others. The "smallest" type is the one obtained simply by counting forever: 0,1,2,3 … and so on. This is *aleph-null* (named after the first symbol in the Hebrew alphabet) and is the first of what Cantor called the transfinite numbers. Such numbers have some decidedly odd properties. For example, adding aleph-null to itself simply produces aleph-null again; so does multiplying aleph-null by itself. This is just the start; Cantor also showed that there are other, even larger, infinities, starting with aleph-one – a number so big that it cannot be reached even by counting for an infinite amount of time. It turns out that there is an infinite number of more infinities, each bigger than the next, until eventually one arrives at the biggest of all, known as Absolute Infinity, denoted Omega. This number is so vast that it is literally indescribable: indeed, its definition is based on the idea that any attempt to describe it can only be describing something smaller.

❓ Is it possible to predict how a tossed coin will land?

In principle, it is indeed possible to predict whether a coin will come up heads or tails. The basic mechanics of coin-flipping are relatively simple and the resulting equations can be solved by computer. A recent study by Professor J P Cusumano and Dr N K Hecht at Pennsylvania State University has shown, however, that the final resting state is only predictable if the coin is tossed so feebly that it barely rises high enough to complete a full rotation. This is because coin-tossing is "chaotic" – that is, even tiny errors in describing the initial state of the coin grow so rapidly over time as to ruin all hope of accurately predicting how it will land. So, for practical purposes, coin-tossing is essentially random. Experimental evidence supporting this was collected by John Kerrich, a mathematician based in Copenhagen, who was interned following the invasion of Denmark in 1940. During his incarceration, Dr Kerrich whiled away the time by tossing a coin 10,000 times and recording the results. He found that the relative rate of heads and tails was entirely consistent with randomness, producing 5,067 heads and 4,933 tails.

? What proportion of heads should raise suspicions about a tossed coin?

England's erstwhile cricket captain Nasser Hussain must have wondered about this during a run of twelve lost tosses in 2001. The odds of this happening by chance are 4,096 to 1 – long odds indeed, but enough to justify suspicions of skulduggery? Faced with such questions, scientists have long relied on a rule which states that something fishy (or "statistically significant" in the jargon) is happening if the odds of getting the observed result by chance alone are less than 1 in 20. For a tossed coin, that means that getting five or more heads (or tails) on the trot should raise suspicions about the fairness of the coin. The bias need not reveal itself as a straight run of heads, of course: if a series of 100 tosses included at least sixty-four heads in any order, that too would be deemed statistically significant. The proportion needed to trigger suspicion moves closer to fifty per cent as the number of tosses increases (reflecting the fact that if the coin is fair, the proportion of heads should also move ever closer to fifty per cent). For 1,000 tosses it is 52.7 per cent, while for 10,000 tosses 50.83 per cent would merit suspicion under the usual scientific rules.

So why, with a run of bad luck over 200 times less likely than the standard scientific criterion for suspicion, did Mr Hussain not cry foul? Because he has more sense than many scientists appear to have. As the tosses occurred in different places at different times with different people, Mr Hussain clearly concluded that the idea of some vast conspiracy having been set up to ensure he lost the toss was far less plausible than the alternative: that he was having a run of bad luck. He therefore just kept going – and in October 2001, started winning tosses again. Unfortunately, many scientists blithely continue to use the 1-in-20 rule to see "significance" in findings that are most likely just meaningless flukes. Statisticians have been warning scientists about this for decades, but few seem to be listening. Perhaps Mr Hussain should go on a tour of the world's scientific laboratories; his approach to probability could save a fortune in wasted research grants.

? How can one visualize a million?

One of the most interesting challenges of science writing is coming up with ways of visualizing things beyond the normal range of experience. Done well, the results can make a lasting impression. I recall reading a book on astronomy which gave a particularly memorable image of the vastness of the solar system. According to the author, if the Sun is represented by a 1 ft beachball, then the Earth is a peppercorn orbiting 30 paces away and Pluto a full stop three-quarters of a mile down the road. As if that weren't mind-boggling enough, on this scale, the nearest star would be over 5,000 miles away.

Illustrating pure numbers of things is a bit harder but one way of visualizing a million is by using graph-paper. A single sheet of A4 graph paper with 2 mm size squares contains around 15,540 squares. By that reckoning, 65 sheets contain just over a million squares. Another possibility is to use sugar: a million sugar grains weigh around 700 g, while a billion will weigh slightly over three-quarters of a tonne.

Such analogies can be especially useful when trying to comprehend the vastness of certain numbers. For example, the chance of winning the UK National Lottery are 1 in 13,983,816, which is one of those numbers that I for one cannot really visualize. Using the graph paper analogy, things become a little clearer: picking the correct six numbers is like choosing a single 2 mm square among a pile of 900 sheets. Alternatively, on the sugar scale, it's equivalent to finding a single black grain among 10 kg of the stuff.

? Is it better to keep the same Lottery numbers, or make a new choice each time?

As the balls emerge from the machines entirely at random and have no memory of what they did before, it make no difference whether you keep the same numbers, change them each week or ask your cat to pick them. At least, mathematically it doesn't make any difference. I would, however, strongly urge against using the same set of numbers each week for the following reason: imagine you have been using the same set of "lucky numbers" for months and finally decide they aren't so lucky after all. Call me Mr Pessimism but I just know that the moment I switched to a new set of numbers, the old ones would hit the jackpot. I, for one, would have problems dealing with that. I therefore recommend that if you really must play this infernal game, then change your numbers each time, selecting fresh ones at random. This won't make the slightest difference to your chances of winning the jackpot but will reduce your chance of having to share your fortune with a lot of other people who all chose the same numbers. Bear in mind though, that research has shown that people are peculiarly hopeless at making truly random choices of numbers, tending to avoid patterns or "runs" of consecutive numbers, despite the fact that these are just as probable as any other selection. So perhaps you'd better get the cat to choose them, after all.

? What are the largest and smallest numbers used in science?

There are various contenders for both these titles. In the 1930s, the Nobel Prize-winning physicist Paul Dirac developed what became known as the "Large Numbers Hypothesis", based around the curious similarity of three numbers: the relative strengths of the electromagnetic and gravitational forces in the hydrogen atom (10^{39}), the age of the universe divided by the time it takes for light to cross a hydrogen atom (also 10^{39}) and the square root of the number of particles in the universe (that is, the square root of 10^{78}, which is 10^{39}. Quite how Dirac ever came to compare these numbers or why they are so similar, remains uncertain. Even so, his work has led to 10^{78} (which is a 1 with 78 zeroes after it) becoming a good candidate for the biggest number with practical relevance in science.

The smallest number of central importance in science has the dubious distinction of being the outcome of the worst-ever back-of-the-envelope estimate. Recent studies of the universe have shown that it is expanding at an ever-increasing rate, propelled by some mysterious "anti-gravity" force. The best bet for the source of this force is empty space itself, which according to the laws of the sub-atomic world is riddled with so-called quantum vacuum energy. It is possible to estimate the size of this energy but the result is way out of kilter with the measured value, which is around 10^{120} times smaller. Explaining why the true value is so small (and yet not zero) is regarded as one of the greatest mysteries of contemporary science.

? Why are mathematicians so interested in prime numbers?

Prime numbers are the building blocks of all numbers greater than 1. That is, every number is either itself a prime, such as 2, 17, 53 or 673, or is the product of primes, such as 17,119 ($17 \times 19 \times 53$). Furthermore, every number can be broken up into its primes in only one way. This is no mere supposition: in 1801 the mathematician Carl Gauss gave a proof of this "Fundamental Theorem of Arithmetic" (though it seems likely that Euclid had a proof of it 2,000 years earlier). Beyond their fundamental nature, primes tantalize mathematicians with properties that seem to be true but defy proof. For example, Euclid himself came up with a wonderfully neat proof that there is an infinite number of primes, yet to this day no one has proved that there is an infinitude of "prime pairs", such as 5 and 7 or 59 and 61, in which two consecutive odd numbers are primes. Then there is *Goldbach's Conjecture*, first pointed out in 1742, which states that every number greater than 5 is a unique sum of just three primes. Again, while this is widely believed to be true, no one has ever succeeded in proving it.

Proving that a given number is prime has long been used to demonstrate calculating prowess. Once performed by "savants" with a gift for mental calculation, the task was among the first given to electronic computers. The current record for the largest known prime is $2^{13466917}-1$, a number of 4,053,946 digits, found by a network of thousands of linked home computers in November 2001.

Since the late 1970s, primes have become of enormous commercial importance, as they form the heart of the RSA encryption system, widely used to protect financial transactions. Roughly speaking, the RSA system is based on the belief that there is no quick way to factorize big numbers into two similarly-sized prime numbers. While many think this is true, yet again a solid proof is lacking. Given what's at stake, I find this rather disconcerting – equivalent to a bank declaring it's pretty sure no one will find the mat under which it has put the keys to the safe.

? Why do people insist on quoting "median" figures instead of just averages?

I'm sure some believe median is just a posh word for average but there is a difference and one which is often of more than merely academic interest. Averages are routinely used to suggest a "typical" value for, say, height, age or salary. This works well enough, especially if the values are fairly evenly spread. If, however, the values are heavily skewed in one direction, it can be woefully misleading. For example, a small firm with six employees earning £5,000 a year and one fat-cat boss on a salary of £110,000 can rightly claim to be paying an average of £20,000. The median casts a rather different light on things by giving the "middle" value – that is, the level at which there are as many people doing worse as there are doing better. In the case of our firm, the median salary is £5,000 – which is just what all but another fat cat can expect to earn.

? Is it true you need just twenty-three people in a room to ensure two have the same birthday?

Only a crowd of 367 people is actually *guaranteed* to have two sharing the same birthday, as it's just possible that even in a crowd of 366, birthdays are spread evenly throughout the 366 days of a leap year. The real "Birthday Paradox" states that only twenty-three people are needed to give *better than 50:50 odds* that two or more share the same birthday. While no longer guaranteeing a match, the small number of people required is undoubtedly surprising (people typically guess that the answer would be around 180). The two key reasons it is far smaller casts light on how easily we can be fooled by "amazing" coincidences.

First, notice that no specific date is mentioned; the paradox requires only a match between any two birthdays. This tripped up a mathematical guest on the *Johnny Carson Show* some years ago. Noting that there were about 120 people in the audience – far more than required by the paradox – Carson decided to put his guest's claim to the test and asked if there was anyone in the audience who shared his birthday, 23 October. There wasn't – leaving the guest looking rather red-faced. By specifying a date, Carson had drastically reduced the chances of getting a match (a crowd of over 250 people is needed to give even 50:50 odds of matching a specific date). The paradox allows any match to count, a far less demanding target.

The second key point is that the number of pairs that can be made from people in a crowd is much larger than you might think. For example, a group of four people can be paired off in six different ways, while ten people permits forty-five pairings. For a crowd of twenty-three, no fewer than 256 pairings are possible. With only 366 possible birth-dates available in a year, it no longer seems so surprising that twenty-three people are enough to give an evens chance of at least one match.

The lack of specificity, plus the pairing effect, helps explain why we are so often surprised by spooky "coincidences". We rarely specify in advance precisely the nature of the coincidence which will surprise us and we often underestimate just how many opportunities for matching are possible from any given collection of objects – be they people or events.

? What are the earliest and latest dates on which Easter can fall?

Easter appears to jump around between late March and late April at random but there is an underlying principle, based on the Biblical accounts of Christ's crucifixion and resurrection. According to the Gospels, the Crucifixion took place on the day after the Last Supper, at or around Passover – whose timing depends in turn on the date of the first full moon after the Vernal Equinox. As if that weren't complex enough, further constraints have been added in the intervening millennia by popes determined to ensure Easter never coincided with the Jewish festival and by the needs of calendar reform.

Over the years, several distinguished mathematicians have devised various means of turning the resulting tangle into a mathematical recipe. Not surprisingly, the formulas are somewhat involved; those wanting the full algebraic details should consult the authoritative *Mapping Time: The Calendar and its History* by E G Richards (Oxford University Press, 1998). Those who prefer to have some of the work done for them will find a simplified recipe in the back of the Book of Common Prayer.

Easter can fall as early as 22 March, though this is exceedingly unusual: it last happened in 1818 and will not occur again until 2285. It can also take place as late as 25 April, as it last did in 1943 and will again in 2038.

As luck would have it, we are approaching a time when we shall be able to experience *almost* the full range of possible Easter dates in very short order: Easter Sunday 2008 will fall on 23 March, before zooming off to almost the other extreme just three years later, landing on 24 April. In Britain, from time to time, these wild variations precipitate outbreaks of the campaign to have Easter fixed to one specific Sunday. In 2003 the UK's National Secular Society called for the provisions of the Easter Act 1928, which permit Easter to be fixed as the Sunday after the second Saturday in April, to be invoked. Yet even this isn't "fixed" in the same sense as, say, Christmas Day – and thus bears no simple connection with the date of the historical event on which the whole thing is based. Reviewing the evidence in *Marking Time* (Wiley, 2000), his fascinating book on all matters calendrical, astronomer Dr. Duncan Steel showed that various Biblical strands (including an apparent reference to a lunar eclipse) point to one most plausible date for the original Easter Sunday: 5 April 33 BCE.

? How do mathematicians calculate pi to billions of decimal places?

Over the centuries, many mathematicians, including Sir Isaac Newton, have whiled away their time calculating ever more digits of pi – though not always with 100 per cent success. In 1873, the mathematician William Shanks held the world record with a hand-calculated value of 707 digits, a monumental effort that took up 15 years of his life: an average rate of one digit a week. Not surprisingly, there was little enthusiasm for checking Shanks's work by repeating it. There was, however, a relatively simple test, based on the widely-held assumption that the digits of pi are entirely random, so each digit, from 0 to 9, should make up around one-tenth of the digits in the calculated value of pi. So, for example, Shanks's value should have had 70 or so 7s. It soon emerged, however, that, towards the end of Shanks's result, there was a curious lack of 7s. Not until 1945 was the cause discovered: Shanks had made a mistake, which resulted in every digit after the 527th being incorrect (or, put more bluntly, he had wasted over 3 years of effort).

Efforts to calculate pi have moved on apace since then; in 2002 Professor Yasumasa Kanada and his co-workers at the University of Tokyo Information Technology Centre announced that they had succeeded in calculating the first trillion – that is, thousand billion – digits of pi. They used a Hitachi SR8000 supercomputer capable of a trillion operations per second; even so, it took the machine twenty-five days to spit out over 1,200,000,000,000 digits. To check the reliability of their result, Kanada and his colleagues performed the titanic calculation twice, using two different formulas. They also carried out the test that had revealed Shanks's blunder and checked on the relative frequency of the digits making up the result. Again, the results agree with expectation: the number of 7s turned out to be 119,999,740,505 – almost exactly one-tenth of the total.

? **Why do the clocks go back about two months before the shortest day of the year but go forward three months after?**

The rigmarole of Daylight Saving Time is often thought to be dictated solely by light levels but if this were true, the change-over dates would be more or less symmetrical about the "winter solstice", on or around 21 December. The lack of symmetry reveals the other reason governments impose the change-over: the need to cut energy by controlling the times we are awake and demanding energy. By making the clocks "spring" forward by an hour in the spring, the sun effectively sets an hour later and so shrinks the time between sunset and bedtime by an hour – cutting the time we are indoors, awake and using energy. Once the nation is on Summer Time, governments like to hang on to it as long as possible and eke out "daylight saving" until the end of October. Since 1996, the whole of the European Union has put its clocks forward on the last Sunday in March and back again on the last Sunday in October. With its different climate and latitude, America starts "daylight saving" a week later, on the first Sunday in April. During the energy crisis of the mid-1970s, the US became so desperate for the energy savings made by daylight saving that Congress ordered daylight saving to begin as early as 6 January.

? Why is the shortest distance between two points a straight line?

Anyone who knows a little about navigation knows it isn't always true: the shortest distance between any two points on the globe (or any sphere) is part of a "great circle", whose diameter equals that of the Equator. It is, however, true on a flat sheet of paper, and a proof can be extracted from theorems about triangles deduced by Euclid in the 4th century BCE. Draw two points A and B, with a straight line between them, then imagine that there is another route connecting them that takes you off this line, via another point, C. Euclid's theorems show that even if C is almost on the original line, its distance from A added to its distance from B is always greater than the original distance between A and B. So, even the slightest deviation away from a straight line always adds a bit of extra distance. It's possible to generalize the idea of a straight line to any curved surface, the generic name for the shortest curve between two points being a "geodesic". Calculating their form was little more than an amusement for mathematicians until 1915, when Einstein came up with his General Theory of Relativity. This showed that light rays follow geodesics, the shape of which provides a measure of the strength of gravity.

? Why is the circle divided into 360 degrees?

Even in the twenty-first century, time and angles continue to be reckoned in the same weird units used by the Sumerians over 4,000 years ago. Quite how it started is still debated by historians but one theory, first put forward 200 years ago, links it to the calendrical "circle" of a year. Sumerian mathematicians spotted that the number of days in a year is close to 360, an integer whose prodigious number of factors makes it easy to chop up into smaller parts. The awkward five days left over were declared public holidays. Another theory argues that it comes from the base-60 arithmetic used by the Sumerians. A circle can easily be divided up into six equal parts by marking off points along its circumference at spacings equal to its radius. If each of these parts is taken to be 60 units, then a full circle will be 360 units – which we now call degrees.

❓ Do goalkeepers stand a chance against a penalty kick?

Whoever decided to put the penalty spot twelve yards from the goal-line when penalties were first introduced into soccer in 1891 knew something about human reactions. It takes around half a second for a typical penalty kick to travel that distance – which just happens to be the typical reaction time of humans. To get some idea of the scale of the challenge confronting a keeper trying to save a penalty, get a coin and hold it at shoulder height. In the time taken for the coin to reach the floor, the keeper is supposed to work out where the ball is going, get to it and stop it.

The upshot is that keepers don't have much chance of saving a penalty unless they can find a way of anticipating where the ball is going to go. Statistics suggest that goalkeepers save just one penalty in five during UEFA football championships – around thirty per cent of which have been decided on penalties.

Studies of the world's best keepers show that they typically make their first move around 100 milliseconds before the ball is struck and base their actions on the position of the legs and feet of the striker. Research by sports scientists has shown that the most valuable clues are to be found in the orientation of the hips and feet. Analysis shows that just before they take the kick, strikers usually plant their non-striking foot in the direction of the final kick. According to Professor Ian Franks and colleagues at the University of British Columbia, this "Watch the Other Foot" rule holds true around 80–85 per cent of the time. Better still, it buys the keeper as much as 200 milliseconds to prepare before the kick is taken. For a ball struck at over 50 mph, that translates into a half a goal-mouth's worth of distance – and the difference between victory and defeat.

? **Why do the dimples on golf balls make them fly further?**

Golf balls acquired their distinctive dimpled pattern around a century ago, when manufacturers took advantage of the discovery that a ball with a rough surface travels further and higher than one that is perfectly smooth. This is decidedly counter-intuitive: surely something with a rough surface will encounter more friction as it flies through the air? The explanation lies in the effect of the dimples on the flow of air around the ball. In flight, a wake of turbulent air forms behind the ball, extracting energy from it, meaning it doesn't travel as far as it might. The dimples give the slightly "sticky" air flowing around the ball something to latch on to and as a result, it wraps itself more smoothly around the surface, reducing the size of the turbulent wake. That, in turn, reduces the energy loss due to drag and allows the ball to travel further for a given whack.

Golf ball manufacturers have spent fortunes trying to find very best combination of size, shape and arrangement of dimples. For many years, relatively simple patterns of around 300 dimples were thought to do the trick. In 1995, the US manufacturer Wilson introduced its 500 Series ball, designed by a NASA aerodynamicist, which has 500 dimples of varying sizes arranged in a pattern of sixty spherical triangles. The large dimples are said to reduce the turbulence effect, boosting lift and maintaining any spin put on the ball, while the small dimples prevent the lift getting out of control. Whatever; they still won't turn a twenty handicap player into Tiger Woods.

? **Where's the best place to buy property on a Monopoly board?**

With hundreds of thousands of people buying this infernal game every Christmas, it is a racing certainty that many will fall into the trap of buying the most expensive property (on a British board Park Lane and Mayfair) because these seem to offer most scope for bankrupting opponents. Which is true – if opponents land on them. But this overlooks the fact that the chances of landing at any given place on the board are not all the same. Part of the reason is that players roll two dice, so that players typically move on by seven places on each throw (there being more ways of getting seven from two dice than any other number).

Mathematicians have looked at the effect of this and other features of the game to find which part of the board is likely to prove most profitable. A 1997 American study concluded that the best pickings begin at Marylebone Station and continue through the red block of Trafalgar Square, Fleet Street and The Strand. However, Dr John Haigh of Sussex University argues strongly for the orange block of Vine Street, Bow Street and Marlborough Street. Frankly, I would always recommend the pub.

? Is it possible that there have ever been two identical games of snooker?

Those who, like me, find televized snooker very therapeutic after a stressful day will have noticed just how often just a tiny difference – perhaps due to a bit of chalk-dust on the baize – can lead to a complete turn-round in a game. This is symptomatic of the phenomenon of *chaos*, in which small deviations lead to large differences in outcome. A rough calculation shows that if two snooker games are played with the first balls struck to within a hair's breadth of the same position, the games will become radically different in fewer than half a dozen shots. Thus it's a fair bet that since the game was (allegedly) invented in India in 1875, no identical games have ever been played.

? How can a ball swerve in mid-flight?

Some years ago, there was much argument among cricket enthusiasts about whether the ball really could swerve on its

95

way towards the batsman. The argument was finally resolved by slow-motion replays confirming that, amazing as it might seem, the effect was entirely genuine. I doubt whether soccer fans ever shared this scepticism, as the effect is routinely use to produce spectacular "banana kicks", swerving round defensive walls.

The basic phenomenon is the *Magnus Effect*, named after the physicist who investigated it around 150 years ago. Unless a ball is kicked dead at its centre, it always spins as it flies through the air, which leads to the relative speed of the air passing over the ball being higher on one side than the other. Most textbooks argue that, as the pressure exerted by any gas is linked to its flow rate, this difference in relative air speed creates reduced pressure on one side of the ball, which is then effectively "pushed" off course by the higher pressure on the other side. Yet, as Dr John Wesson pointed out in his recent study of the phenomenon, this argument makes various assumptions whose validity is hard to assess. He suggests that a more plausible explanation is that the effect exploits the viscosity or "stickiness" of the air, plus the presence of nooks and crannies on the ball's surface. This leads to the ball behaving rather like a water-wheel, with air molecules clinging to the ball before being hurled off one side at relatively high speed, the resulting thrust making the ball swerve in response.

The effect can be truly spectacular, as the Brazilian soccer player Roberto Carlos showed in a match against France in June 1997. He sliced the ball with the outside of his left foot from 35 m out from the goal, sending it spinning, at 85 mph, to the right of the French defensive wall. The ball's trajectory took it so far away from the goal-mouth that a ballboy standing on the goal-line ducked to avoid what he thought was a certain hit. Then, as the ball slowed, the magic of the "banana kick" began to reveal itself, pushing the ball into an ever-tighter curve – and the back of the net.

❓ Which casino game offers the best chances of winning?

Winning is only half the problem; there's also the issue of hanging on to your winnings. Casinos make money through the "house edge", the difference between the true odds of a win, calculated from probability theory and the slightly worse odds offered by the house. In roulette, for example, the odds of picking the correct number are 36 to 1, while the casino offers 35 to 1, giving a house edge of about three per cent.. Some bets – like gambling on whether the ball will end up in red or black – carry a slimmer house edge but in the end, the laws of probability mean that the casino will come out on top. The only hope in these cases is to detect some bias in the mechanics of the game, such as a slightly faulty wheel, which is big enough to swamp the small house edge. Naturally, casinos take great pains to eliminate such bias. Many professional gamblers focus on the card game, blackjack. In 1962, the mathematician Dr Edward Thorp showed that blackjack can give a slightly negative house edge, favouring the player. Beating the casino involves tracking the changing composition of the decks and making the appropriate bets. With over 300 cards to memorize in a standard six-pack shoe, this is no mean feat – and casinos are always on the alert for glassy-eyed punters who look as if they're "card counting".

? What is the reason for the pattern of pentagons and hexagons that make up the surface of a soccer ball?

One doesn't normally associate soccer with higher mathematics but the patchwork of twelve black pentagons and twenty white hexagons covering the traditional soccerball is a reflection of two deep theorems. The first, proved by the great mathematician Carl Gauss in 1828, shows that it is impossible to make a perfect sphere out of anything flat: there will always be some wrinkles or distortion. Map-makers knew this long before Gauss formally proved it, and had come up with a whole range of map "projections", which capture the curved surface of the Earth on flat sheets of paper but always with distortions of one type or another.

For manufacturers, the problem is how to create a robust, inflatable ball that is as close to a sphere as possible. A patchwork of fairly large panels is a good solution, as it minimizes the amount of stitching, copes with the impact of kicks and bounces fairly well. The question of what shapes the panels need to be is answered by a theorem found by Rene Descartes in 1635 (but named Euler's Formula, after the mathematician who rediscovered it over a century later). This states that, for any solid made from polygons, the sum of the number of faces and corners equals the number of edges plus 2. For a cube, for example, we have 6 faces and 8 corners, making 14, which equals the number of edges – 12 – plus 2. Using this formula, we discover that a combination of twelve pentagons and twenty hexagons is needed to create the simplest fairly spherical solid. It is known as a truncated icosahedron and can be found bouncing around pitches across the world.

❓ What is the origin of the "home advantage" in soccer?

There's no doubt that the home advantage is very powerful in soccer. A monumental study of over 4,000 matches played by English league clubs during the 1990s, performed by Dr Mark Dixon and Dr Michael Robinson at City University, London, showed that almost a half of home games produced wins compared to around a quarter of away games. Not surprisingly, there has been extensive research into the causes of home advantage, with everything from travel stress to referee bias being investigated. Overall, psychological factors seem to be the principal cause. A study of the size of the advantage in different types of matches, by Dr Richard Pollard of the University of the South Pacific, has shown that it is relatively small in local games and relatively big in international matches – as one might expect if away players were being influenced by feelings of alienation. Some backing for this comes from research showing that games like baseball, where managers have plenty of opportunity to rally intimidated players, have a much smaller home advantage than soccer, where away players are compelled to face a roaring home crowd on their own for forty-five minutes at a stretch. Before one condemns the players' lack of mettle, it is worth noting that match officials have also been shown to be intimidated by the sheer racket created by tens of thousands of fans. In 1999, *The Lancet* carried the results of a study by Dr Alan Nevill and his colleagues at Liverpool John Moores University which asked subjects to judge fouls video-taped at a real match. They found that if the sound was left on, those watching were more likely to judge away players guilty of fouls. If the sound was turned off, this bias vanished: clear evidence that, in soccer, who yells wins.

? Is there a reason why there are so many incorrect offside decisions in soccer?

The offside rule causes more controversy than any other. At its simplest, the rule states that a player is offside if, at the moment the ball is passed by a team-mate, there is only the goalkeeper between them and the goal-line; there are various tweaks about throw-ins and the like but this is the essence. It also captures the nub of the problem: match officials have to judge the relative locations of the ball, the attacker and the last defender, all in different positions and all at the same time. It's not easy, and research suggests that they get it wrong around twenty per cent of the time.

More training would seem the obvious answer but a recent study by scientists at the Free University of Amsterdam, Holland, puts the blame on an optical illusion which leads to attackers being deemed offside more often than they really are. To see how it works, hold a pen upright in each hand and line them up at arm's length, so that one is hidden behind the other. If the pen furthest away were an attacker who had just been passed the ball, they would not be offside. Now move your head to the right and look again: the "attacker" now seems to be ahead of the defender and so offside. Thus, any assistant referee looking back towards the players, with the attacker furthest away, is in danger of incorrectly flagging offside. Sure enough, the scientists found that the assistant referees look backwards around ninety per cent of the time.

Chapter 5

Matters meteorological

? **When caught in the rain, is it better to run or walk?**

This is one of those questions that confirms the maxim of the Nobel Prize-winning physicist Philip Anderson: there is no problem which, looked at in the right way, does not become even more baffling. On the face of it, the answer seems obvious: running must always be preferable, as it minimizes the amount of time spent in the rain. Then the doubts start to creep in. Every driver knows that the faster one goes during a shower, the greater the amount of rain hitting the windscreen per second and the faster the wipers must work to keep it clear. So, could it be that if we walked through the rain, we would encounter less rain per unit time and thus end up drier? Cranking through the maths, it turns out that this effect is secondary: you are always drier if you run. Then comes the final twist: using real figures for rainfall rates, wind and running speeds, it turns out that the benefits of running, while real, are decidedly marginal. My advice is that if caught in light to moderate rain, running for shelter is not worth the effort unless a downpour seems imminent.

? When is the hottest part of the day?

Being English, I am tempted to say "About ten minutes after you put the hot water bottle under the duvet". Noel Coward gave a plausible answer, in his famous 1931 study of foreigners, mad dogs and Englishmen, where he pointed out the failure of the latter to accept the dangers of going out at noon, when the sun is at its highest in the sky. Yet, while what Coward called the "ultry violet" rays are indeed most intense then, the temperature has some way to go. The ground continues to mop up the sun's heat, steadily raising the temperature of the air above it for several hours after midday. Sean Clarke of the UK Meteorological Office tells me that on a still, clear summer's day peak temperatures are typically reached around 4 p.m. (and occasionally even later) after which the sun's declining altitude in the sky allows things to cool off.

? Why can you occasionally see more than one rainbow?

A rainbow is the result of raindrops ahead of us being struck by sunlight from behind us, which is then bent and separ-ated into its component colours. The rays of coloured light emerging from each raindrop after one bounce are bent through around $42°$, so we only see rays from those raindrops that just happen to be in such a position in the sky that the $42°$ bending effect sends them straight into our eyes. All the rays meeting this criterion lie on a circle; the part of that circle not cut off by the horizon is what we call the rainbow. If the sunlight is bright enough, it's possible to see a second, fainter, rainbow outside the first, the result of the sunlight bouncing a second time inside the raindrop before emerging. And a third, even bigger rainbow? Certainly it seems reasonable enough, yet as the seventeenth century scientist Edmond Halley first proved, the third rainbow should appear around the sun itself. It should also be even fainter than the second rainbow, which may well explain why no one appears ever to have seen it.

The colours of the rainbow appear because as the beam of light passes through the raindrop each colour within it is bent (refracted) by a slightly different amount, depending on its wavelength (red being refracted less than violet), resulting in the originally white sunlight being split into all the colours of the rainbow. But why do we see one single rainbow, rather than millions of tiny ones from each individual raindrop? This is because we only see the rainbows produced by drops located in the exact position from which their light can enter our eyes – in short, we are each supplied with our own, personalized, rainbow. If the second, larger rainbow is also on view, notice how the colours are the reverse of the main bow – the second reflection leads to the order of the emerging colours being flipped over.

❓ Is it possible to drive through the end of a rainbow?

Rainbows are not so much objects as relationships involving the positions of the sun, clouds of water droplets and ourselves. They appear only when our eyes are in a position to intercept the coloured beams of light produced by sunlight being refracted and reflected from the droplets. The laws of physics prevent all but a fraction of the beams of sunlight meeting this demand at any one time but the sheer size of a rain cloud usually allows us to travel some distance before running out of suitably placed raindrops.

However, there will come a point when the rays cannot enter our eyes any more and our rainbow will disappear, even though it may still be visible to others. One of these locations is at the end of a rainbow: if you are standing there, the required relationship between the sun, the cloud and yourself no longer holds and whatever else you might see, it's not part of the original rainbow. This is not to say that reports of driving through rainbows are figments of the imagination. The key phrase is "driving through": this suggests that what people see when they think they are passing through a rainbow is the result of sunlight refracted from drops very close to their vehicle – perhaps in the form of spray thrown up from the road. Suitably positioned, such spray will produce a spectacular display of colours that appears to be part of the huge rainbow hanging in the sky. You would thus have a vivid sense of having driven through the rainbow, yet when you look in the rear view mirror, the illusory nature of the effect becomes clear – there is nothing there at all.

❓ Why do the lights of a distant town sometimes twinkle?

The twinkling effect is the result of turbulence caused by heat changing the density and thus the optical properties of the intervening air. Heat rising from buildings in towns ensures the air is always turbulent around them but we don't usually notice its effect on the appearance of nearby sources of light because it is small, relative to the apparent size of the objects. On the other hand, lights from a distant town appear so small that the effect has a significant impact, which we see as twinkling. The same phenomenon, incidentally, allows one to tell the difference between stars and planets in the night sky. The point-like images of the immensely distant stars are affected by turbulent air far more than the planets, which appear to be relatively twinkle-free.

❓ How much rain does an inch of snow equate to?

This seems a simple enough question: just a matter of looking up the density of snow relative to that of rainwater and multiplying it by the depth to get the equivalent amount of rainfall. The problem is that there are many different types of snow: a study conducted in the mid-1960s at Hokkaido University, Japan, identified no fewer than eighty varieties, ranging from simple needles to the classic ornate six-sided flake. They also pack together in different ways, giving different snow densities, which equate to different levels of rainfall. For the fluffiest – lowest density – snow the conversion factor is around 50 to 1; that is, one inch of snow equates to just 1/50th of an inch of rain. On the other hand, snow crystals that warm up as they descend to earth produce a much denser covering, and a conversion factor of just 4 to 1. As a rough rule of thumb, a figure of around 7 to 1 seems reasonable, so that two inches of snow amounts to around 1/4 inch of rain.

? Why does it seem so quiet after a fall of snow?

Even city streets become a little more tranquil after snowfall – at least at night. The reason is that the top layer of freshly-fallen snow is made up of spiky crystals stacked on top of one another, with nothing but air between them, making them almost foam-like. When struck by sound waves from sources of noise, they mop up most of the acoustic energy, returning hardly any at all. Hard roads, cars and roofs covered with this sound-absorbing material results in relative tranquillity – until the kids come out to play.

? Is there any basis to the weather folklore about it being "too cold for snow"?

I've become very wary of questions about the weather, as the "obvious" answer all too often falls foul of some more subtle feature of the way the weather works: and thus it proves here. As snow is made from ice crystals, which form at any temperature below the freezing point of water, the idea that the temperature can be too low for snow to form seems to contradict elementary science. Yet this overlooks the fact that the temperature is pretty irrelevant if there's not enough moisture in the air to form ice crystals – and bitterly cold weather is often the result of relatively moisture-free high-pressure systems. So, during a cold snap it can indeed be "too cold for snow" and we must wait for a warmer and wetter low-pressure system to arrive. If one follows hard on the heels of a cold snap, there is a much better chance of the snow settling, as otherwise it takes some days for the ground to cool sufficiently to prevent the snow crystals melting on contact.

? Is it really true that each snowflake is totally unique?

The shape of a snowflake is a result of water molecules sticking around a speck of atmospheric dust and turning into ice. As this is an essentially random process and takes place in ever-changing temperature and vapour levels, the chance of any two snowflakes being precisely the same is very slim. That said, all snowflakes fall into forty or so basic shapes, ranging from long, thin needles to flat plates. Most, but by no means all, feature the familiar six-sided symmetry of the traditional snowflake, which reflects the hexagonal arrangement of water molecules. Simpler-looking shapes clearly have a higher chance of looking alike and in 1988 the *Bulletin of the American Meteorological Society* reported the discovery of two very similar rectangular snowflakes on a research plane flying over Wisconsin in November 1986. As far as I am aware, this is the closest that anyone has ever come to finding two identical snowflakes.

? Do Eskimos really have eighty different words for snow?

It seems entirely reasonable that Eskimos should have many words for snow; after all, it is what they spend most of their days looking at. Figures quoted in the literature range from around half a dozen to 400-plus but most are, however, the purest nonsense. The trouble started around 1940, when Benjamin Lee Whorf, an amateur linguist, declared that Eskimos recognize at least seven different types of snow. This claim then underwent the Chinese Whisper process until it reached ludicrous proportions. Linguists point out that

Eskimos do indeed have a large variety of snow-related words but they do not refer to different types of snow. They total no more than about a dozen – similar to the number recognized in English.

❓ How do weather forecasters estimate wind-chill?

The first attempt to quantify the chilling effect of wind was made in 1939, in the doctoral thesis of the geographer and scientist Paul Siple. Working with Charles Passel, he conducted a series of experiments in Antarctica to find out how temperature and wind speed combined to affect the rate of heat loss from the human body – or, rather, from plastic cylinders of water hung out in the Antarctic winds. Despite being somewhat academic, the findings were deemed of value to military planners and kept secret until after the Second World War. They were also rather hard to interpret, as the chilling effect was quoted in units representing energy loss per square metre of area – not exactly a concept in common currency. In 1973, the US National Weather Service took the obvious step of quoting wind-chill in terms of equivalent temperature: that is, the temperature one would feel when exposed to such conditions, for example, a day on which the thermometer showed 0 °C when there was a 20 mph breeze blowing would feel more like –14 °C.

Inevitably, making the conversion from Siple's results to something more comprehensible produced a complex formula, with terms based on wind speed, ambient temperature and a combination of the two. It also proved pretty unreliable as a guide to just how cold it would feel – not least because of the unrealistic nature of Siple and Passel's original experiments. Over the last few years, this has prompted renewed attempts to capture the concept of wind-chill and in 2001 the US National Weather Service adopted a new formula based on experiment and computer simulations. This gives less dramatic wind-chill levels, with a 20 mph breeze making a 0 °C day feel around –7 °C – considerably warmer than before.

Some meteorologists think this revision does not go far enough and argue that wind-chill depends on a host of factors, such as humidity, altitude and cloud cover, as well as on temperature and wind-speed. None of this impresses the British Meteorological Office, which has long been sceptical of the whole concept of wind-chill. When it does make reference to it, it uses a relatively simple formula devised in 1971 by R G Steadman, which gives values between those of the old Siple formula and the US NWS. Or, as they more simply put it: "It will feel much colder in the biting wind".

? In weather forecasts we hear of high and low pressure – but is there a medium pressure in between?

There is a dividing line between high and low pressure, set at a pressure of – surprise, surprise – one atmosphere. Its actual value is set according to a standard "reference model" of the atmosphere arrived at in 1965, according to which sea level pressure is 1013.25 millibars. A pressure above this counts as "high pressure", while one below is "low pressure". On 31 January 1902, a British record high pressure of 1054.7 millibars was registered at Aberdeen; the world record high pressure is 1083.8 millibars, registered at Agata, Siberia, on 31 December 1968.

? If the weather is unpredictable, can we believe climate change predictions?

According to meteorologists, weather forecasts more than about twenty days ahead are ruled out by the "chaotic" nature of the atmosphere, under which tiny errors in the original data will come, over time, to overwhelm the forecast. So, what faith can we have in forecasts of what the Earth's climate will be like decades from now? For years the stock answer has been that climate prediction focuses on broad-brush factors, such as summer temperatures and winter rainfall and these are less sensitive to the effects of chaos. Yet, recently, I have begun to detect growing concern that this might be a bit complacent and that climate forecasts may well suffer significantly from chaotic effects.

This is what lay behind the call from scientists in September 2003 for anyone with a decent home computer to help in a huge climate modelling project. The idea is to run many thousands of simulations of the Earth's climate, each with slightly different starting data and see how the final results differ. This "ensemble forecast" method is widely used to gauge the effect of chaos on weather forecasts. Now it is being applied to climate modelling. It will be interesting to see just how big an effect chaos does have on climate models. I for one won't take any of the doom-laden scenarios seriously until researchers start putting some error bars on their predictions.

? How are hurricanes named?

Once identified, hurricanes are allotted names from a list drawn up by the US Weather Service. For many years all the names were female but in 1979 protests from feminists led to these violent tropical storms being allotted an equal proportion of male names. (Apparently there are now calls for more Afro-American names to be used. Don't ask.) The list is re-used every few years, so the same names come round again – apart from those that turned out to represent truly horrendous hurricanes, like Hugo of 1989 and Andrew of 1992. These are granted meteorological immortality.

? When forecasters talk of a 90 per cent chance of rain, what do they mean?

The figure is meant to give some measure of the probability of it raining, usually within a specific area and time. The trouble is that the very meaning of probability is far from cut and dried and has long been a source of argument. According to one school of thought, probability can be thought of as a statement of the frequency with which an event can be expected to occur. For example, in the long run, half of all coin tosses will produce "heads", so the probability of a "head" is said to be 1/2. Seen in this light, a 90 per cent chance of rain means that if you took your umbrella every time the forecasters made such a prediction, they think that in the long run you would have done the right thing 9 times out of 10. There is, however, another school of thought, which holds that this probability is really a statement of the level of belief of the forecasters – 90 per cent implies a great deal of confidence. Which of these interpretations are forecasters actually using? Ewen McCallum, head of forecasting at the UK Meteorological Office, tells me it is a judicious mix of both.

To gauge the reliability of each forecast, the Meteorological Office runs its computer model around fifty times, altering the starting conditions slightly each time. If forty-five of the runs predict rain over The Oval at lunchtime, then forecasters can warn cricket fans of a 90 per cent risk of rain then. Meteorology is, however, a subject which still requires human skill and judgement and the probability figure also often reflects the level of confidence the forecasters have in the computer's predictions. So why not simply say "a serious risk of rain"? The use of a specific figure helps reduce the vagueness of such statements and, according to Mr McCallum, is especially useful to businesses, local authorities and others who need hard figures to plug into decision-making software. For the rest of us, Mr McCallum offers a rule of thumb: respond to probabilities of rain like traffic lights: values over 70 per cent are "red", those between 50 and 70 per cent "amber" and anything below 50 per cent "green".

? What is responsible for the deep blue colour of really pure ice?

Snow usually looks white because its surface scatters and reflects equally all the colours making up sunlight. If light is able to penetrate some way below the surface, another effect becomes important. Water molecules have a slight preference for absorbing red and yellow light, boosting the relative amount of blue in the reflected light. This effect only becomes noticeable in relatively deep amounts of water – hence "the deep blue sea". In the case of snow and ice, light emerging from holes or crevasses a few metres deep has undergone enough absorption of its red and yellow components to leave it looking decidedly blue.

? What is the difference between celsius and centigrade?

In my experience, the former is the preferred temperature scale of engineers and tiresome retired schoolteachers. In reality, there is no difference at all: ice melts at 0°celsius or centigrade, and water boils at 100° of either too. The reason we ended up with temperatures named after an obscure eighteenth century Swedish astronomer is almost too daft for words: in 1948 an international team of experts decided we were in danger of mixing up the "centigrade" used for temperatures with one-hundredth of a "grade" – a measure of angle which I suspect most people have never heard of, much less mixed up with temperature.

? Is there any truth in "red sky in the morning, shepherd's warning"?

This is a venerable bit of meteorological lore (according to Matthew 16, even Christ knew of it) and weather expert Paul Marriott confirms it to be around 70 per cent reliable. The red sky is the result of early sunlight scattering off particles in low-level clouds – which in high latitudes often presages bad weather to come.

? Does sound travel further during hot weather?

Yes, especially in stickily hot weather. Sound waves are formed whenever a column of air – or indeed anything else – is alternately squeezed and stretched. It takes a certain amount of energy to set the air in motion but as it spreads out the sound wave eventually runs out of puff. Just how far it gets depends on how much energy it starts out with and how quickly this seeps away, which in turn depends on atmospheric conditions.

As is so often the case with the physics of a fluid such as air, the details are complex and often counter-intuitive but the upshot is simple enough: sound waves travel much better when it's really humid. The effect is dramatic, especially on higher-frequency sounds, which travel much further (or, equivalently, seem to come from much nearer by) when the relative humidity climbs into the 90 per cent-plus region. Noises which you might normally only hear as a dull rumble become very clear, especially from individual sources such as a car or light aircraft.

According to the UK Meteorological Office, on most days the relative humidity falls from a peak of around 95 per cent near dawn. This may well explain why traffic and birds can seem so damnably noisy first thing in the morning.

**? How does modern rail track cope with hot
weather without expansion joints?**

After the speed restrictions imposed on the UK train network in 2003 to prevent track buckling, it's tempting to say "It doesn't". Yet in the summer of 2003, Britain had record-breaking temperatures, where some parts of the track reached levels of 52°C. Given that the track also has to cope with sub-zero winter temperatures, the wonder is that the network can handle so huge a temperature range so well.

Old-style track, which makes up about a quarter of the UK network, has a gap around 3/8ths of an inch wide between each 40-ft length of track, which mops up the expansion of the track as it heats up. That can cope with a temperature range of around 32°C. New-style track, known as continuously welded rail (CWR), uses a clever trick in which the steel track is stretched as it is laid down and welded into place. The stretching has the effect of putting more space between the atoms in the steel, giving them more room to move as the temperature builds up – effectively reducing the thermal expansion rate of the steel. There is a limit to how much pre-stretching can be done, as it reduces the cold-weather performance of the steel but gener- ally, pre-stretched CWR performs better than the old track, with its clumsy expansion-joint method.

Chapter 6

The natural world

? **Can animals really sense an impending earthquake?**

A few days after the Indian Ocean tsunami struck on 26 December 2004, reports began to emerge of strangely low levels of fatalities among wildlife. According to officials at the Yala National Park in Sri Lanka, where sixty humans lost their lives, not a single dead animal was found – despite it being one of the worst-hit areas. Most striking of all were the reports from India's Cuddalore coast, where thousands of humans died, while buffalo, goats and other animals seemed to have escaped unscathed.

I must admit that when I first heard this I suspected it was simply one of those myths which sometimes emerge at such times, like "signs and portents" in the sky. Yet the early reports have been confirmed by many witnesses, who reported seeing elephants running for higher ground, flamingos leaving their low-lying breeding grounds and other strange behaviours.

While these tie in with accounts following many other major seismic events, sceptics rightly object that they may all be *post hoc* rationalizations. After all, spend enough time at your local duck pond, and you'll witness "strange behaviour" just before, for example, a bomb attack in Baghdad. Some scientists have also pointed out that the intervals between major earthquakes are too long to allow animals to associate what they are sensing with memories of past disasters or give much scope for evolution to promote the emergence of quake-sensing traits.

Such arguments are far from compelling: perhaps the animals are benefiting from sensory abilities acquired for other reasons, such as communication. In 1997, researchers at the University of California reported that elephants could detect the stomping of others over distances of more than 30 miles. Or maybe the animals were responding to the faint seismic effects that often presage major quakes; there were plenty of those in the run up to the Sumatra quake – the area had been struck by three of the seven most powerful quakes recorded anywhere in the world in 2004.

Alternatively, they might have sensed electromagnetic disturbances that some scientists claim accompany the fracturing of rock before a quake. In 1998, a team of scientists in Japan examined this possibility by watching the behaviour of laboratory animals as blocks of granite were mechanically crushed nearby. As the pressure on the blocks mounted, the animals became increasingly nervous and this appeared to be linked to the emergence of electromagnetic effects from the crushed rocks. The team suggested that this may explain how some animals seem able to sense major quakes hours or even days ahead. The terrible events of December 2004 should surely prompt more research into this fascinating possibility.

? Why are there approximately as many men as women?

If men were as valuable as women reproductively, this wouldn't be much of a conundrum but the fact is that men just aren't as reproductively precious, being able to create huge numbers of offspring in the time a woman takes to get through a single pregnancy. So why not a sex ratio of, say, 1 man to 50 women? The most widely-accepted explanation was given by the great geneticist Ronald Fisher in 1930: suppose the sex ratio really was 1 man to 50 women; that man would have lots of opportunities to mate and in the process pass on any genes that favoured the fact he was born male rather than female (and such genes do exist). As his offspring increased in number, so would the prevalence of those genes and hence the proportion of males, until it reached a 50:50 ratio. Then the very deficit that gave males a breeding advantage would vanish and the ratio would stay at 50:50. The same argument works for females, thus keeping the sex ratio pretty much fixed over time.

At least, it would if only genes were at work. In the real world, we have a host of other factors, such as the relative costs of raising sons and daughters and their long-term economic value. This has led to the grotesque practice of female infanticide, which in China is thought to have contributed to a twenty per cent excess of baby boys over girls.

In the world as a whole, there are around five per cent more males than females but the UK sex ratio is currently the other way round, with a five per cent deficit of males. Quite why is unclear.

? Why do humans walk upright?

This is one of those mysteries that just refuses to lie down. At the last count, over a dozen theories had been put forward over the last century or so. In 1871, Charles Darwin made the perfectly reasonable suggestion that bipedalism emerged in order to free up the hands for holding tools. While this would no doubt have brought an evolutionary advantage, it does not seem to have been the primary driver: fossil and molecular evidence suggests that the first humans were standing upright around seven million years ago – 4.5 million years before they began making tools. In the 1950s, the anthropologist Raymond Dart pointed out that standing upright allows creatures to see further – which is true, though even a doubling in height increases visual range by less than fifty per cent. A more recent suggestion is that bipedalism reduces the skin surface area exposed to the noonday sun; again, calculations suggest the benefits are marginal. Perhaps the most comprehensive explanation was put forward in 1980 by Peter Rodman and Henry McHenry of the University of California. They argued that climate change, which led to dwindling woodland conditions and thus more sparse food resources led to bipedalism, as it is a more energy-efficient way of getting the required amount of food. Critics immediately pointed out that studies had shown that quadrupeds (like dogs) get around far more energy-efficiently than humans. Which is true, but also irrelevant: the point is that a bipedal human is more energy-efficient than a human forced to get around on all fours. This seems pretty cogent to me but it hasn't stopped others coming up with yet more theories. In his book *Lowly Origin* (Princeton University Press, 2004), Dr Jonathan Kingdon of the University of Oxford suggests that "squat-foraging" with the hands led to changes in anatomy that made bipedalism virtually inevitable – a neat reversal of Darwin's original suggestion.

? How can you tell the difference between alligators and crocodiles?

While they may look pretty much the same, these two creatures actually belong to two quite different taxonomic families. To biologists, they are as distinct as humans are from gor-illas. One pretty reliable way of telling the two apart is to look at their heads: alligators have U-shaped snouts, while crocodiles' are V-shaped (easy to remember: A comes before C in the alphabet and U comes before V).

This also leads to another rule: the wider jaw of alligators means the lower row of teeth is hidden, while a crocodile shows a set of interlocking teeth, plus a characteristic protruding fourth tooth (again, easily remembered via "crocs show canines").

On the whole, alligators tend to be larger and less aggressive than crocodiles. However, avoiding any contact with either of them in the wild is 100 per cent reliable advice.

question is that we are the same as we were ten years ago: to wit, nothing but convenient fictions.

? Why do men have nipples?

This is one of those questions that begs for a Darwinian "Just-So" story, where once upon a time men and women both breast-fed, until a random mutation produced males unable to breast-feed yet better at fighting predators because of a decent night's sleep, leaving the nipples as the only evidence of their former ability. There may well be an evolutionary connection somewhere but it lies well before the advent of modern humans, as the events leading to nipples in men take place very early in pregnancy, suggesting primordial causes. Around a month or so after conception, a group of cells in the embryo linked to sweat glands begin to form the breast and nipples. This is around eight weeks before the embryo is exposed to the hormones, like testosterone, that lead it to develop into a male baby – by which time the nipples have already been formed. Nor is it the only link with our androgynous past: males have more breast tissue that one might think, and each year, in the UK, around 250 men develop breast cancer, resulting in around 100 deaths.

? What is the memory capacity of the human brain?

Attempts to estimate the storage capacity of the 1.5 kg of squidgy stuff sitting in our skulls date back half a century, when the development of electronic computers prompted comparisons between their relatively feeble memories and the (presumably) far larger capacity of the human brain. At a lecture given at Yale University in 1956, the computer scientist John von Neumann came up with an estimate of thirty-five million million megabytes – colossal, even by the standards of today's computers. Later estimates were based on assumptions about the information storage capacity of the 100 million million connections between brain

? Can ships really be sunk by giant squid?

When I was a child, I came across an engraving of a galleon being attacked by a squid and it gave me nightmares for days. The yachtsman Olivier de Kersauson probably still breaks out in cold sweats when he thinks about an encounter he and his crew had in January 2003. While trying to break the world circumnavigation speed record, their huge, 110 ft, trimaran suddenly came to a dead stop one night in the middle of the Atlantic. Peering through a porthole into the sea below, one of the crew was horrified to see a huge tentacle, wider than a human leg, grabbing at the yacht's rudder. The entire vessel began to pitch and shudder as the creature tried to get a better grip. Fortunately, just as it seemed on the brink of breaking it off, it gave up and disappeared. One crew member estimated the whole creature was almost ten metres long.

The response of scientists seems barely more enthusiastic than for reports of UFOs. I can't imagine why, as there is no reason why such monsters could not exist in the food-rich, neutral buoyancy environment of the sea and there is mounting evidence that they do. In April 2003, Dr Steve O'Shea of Auckland University of Technology in New Zealand unveiled the remnants of a squid caught while attacking a trawler's nets off Antarctica. It is thought to be a juvenile *Mesonychoteuthis hamiltoni*, a super-squid first tentatively identified in 1925. According to Dr O'Shea, its dimensions implied that the fully-grown adult could have a total length of almost fifteen metres. As this is similar to the basic dimensions of galleons of the type in which Sir Francis Drake circumnavigated the world, I am beginning to wonder if that engraving was so far-fetched after all.

? If telepathy is possible, why hasn't evolution made it commonplace?

This is a neat argument often put forward by sceptics of "paranormal" phenomena such as telepathy; it was first raised,

I think, by the science fiction writer Isaac Asimov. Given the evolutionary advantage a "sixth sense" would give organisms in finding mates or prey, it is surely remarkable that no animal has developed such abilities. Leaving aside for the moment the possibility that some actually have, there are several rejoinders to what we might call *Asimov's Paradox*. First, just because an evolutionarily useful effect exists does not guarantee that creatures must exploit it. Superconductivity, which allows electric current to flow without resistance, may well be useful for some living creatures. The trouble is, this phenomenon only reveals itself at temperatures below –77 °C and it is hard to see how random mutation and natural selection will ever produce a creature that can cope with those conditions. Similarly, we still await the emergence of badgers with X-ray vision, fusion-powered fish or atomic roosters.

Some would argue that the most serious flaw in Asimov's Paradox is its assertion that there really are no creatures with paranormal powers. Over the years, there have been various claims for psychic parrots and dogs who sense when their owners are coming home. While intriguing – and certainly entertaining – the evidence is far from conclusive. I am personally more convinced by the evidence for the existence of telepathic abilities in humans, as apparently demonstrated in experiments by researchers in well-respected academic institutions such as the University of Edinburgh.

These experiments suggest that while telepathic communication between humans may exist, the effect is rather weak – like trying to pick up Radio Brisbane on a crystal set. This suggests that the reason we don't communicate via telepathy is that evolutionary processes have led us down the more efficient route of exploiting the far stronger emissions of sight and sound that surround us. That said, one cannot rule out the possibility that there are creatures that make routine use of telepathic powers. The best place to look for such organisms would be in an environment where the five conventional senses don't work very well, such as the abyssal depths of the world's oceans.

? In terms of the cells in our body, are we really not who we were ten years ago?

The cells that make up our bodies are constantly being replaced or repaired, their lifespan usually bearing some relationship to their function. Disease-fighting white blood cells may last for only for a few hours on "active duty", while stomach and colon cells have spans of a few days and red blood cells around four months. Skin cells last about a month, while our skeleton is completely regenerated after around 10 years. So, during a typical lifetime, humans get through around 900 skins and about half a dozen skeletons. The major exception to this replacement process is nerve cells, including those that make up our brain and thus our mental facilities. Barring accidents, degenerative illness and booze, nerve cells typically last as long as their owner.

This brings us back to the question of whether anyone is really the same person they were ten years ago. In terms of many of the cells from which we are constructed, the answer is patently no and one glance in the mirror will show that our outward appearance is also very (and usually depressingly) different. Yet we all have a sense that none of this defines our true being; as far as I know, it is no defence in law against prosecution for crimes committed years ago that at a cellular level we are not the person we once were. It is to the contents of our skulls that we tend to look for the essence of being, and the neurons found there are essentially the ones we were born with. What does change is the pattern of connections between them and the way they respond to stimuli. Yet it would be a brave neuroscientist who would point a finger at a specific set of neurons and declare them to represent the essence of our being. The problem of pin-pointing the real "I" has been noted by some distinguished philosophers, the Buddha and Hume among them, prompting them to doubt whether it even makes sense to talk about the existence of a "self". I must admit to finding this line of argument pretty compelling and to that extent I suggest that the answer to the

cells and led to figures of around forty million megabytes; huge but nothing like as big as von Neumann's figure.

Reviewing the various guesstimates, the computer scientist Professor Ralph Merkle of Georgia Tech, Atlanta, notes that they all make more or less implausible assumptions about how the brain stores information. He argues that better estimates are likely to come from experiments which attempt to measure the storage capacity more directly. As an example, he cites the results of studies carried out in the mid-1980s by Thomas Landauer and his colleagues at Bell Labs, New Jersey, in which people were asked to recall different forms of information over varying amounts of time. The level of recall turned out to be surprisingly consistent and, when converted to a value for the total amount of data a brain could store over a lifetime, produced a figure of around 200 megabytes. This is no less impressive than von Neumann's figure but for precisely the opposite reason: PCs typically have far more storage capacity than this. As Merkle puts it: "While this might come as a blow to our egos, it suggests that we could build a device with the skills and abilities of a human being with little more hardware than we now have – if only we knew the correct way to organize that hardware".

? How can birds sit on power lines without getting electrocuted?

It's not high voltage *per se* that is dangerous but high voltage differences, as these produce a flow of electricity from one point to another. It's like walking around on the 100th floor of a skyscraper; the sheer height presents no danger in itself. Only if you open yourself up to the effects of a big difference in height – by, say, jumping out the window – are you in trouble. When a bird lands on a power line, it acquires the same voltage as the cable and so there's no risk of electrocution. If, however, it spread its wings and touched something at a different voltage, for example another cable or some point connected to earth, the voltage difference allows electricity to flow, frying the bird instantly.

Power companies have devised ways of spacing the cables or otherwise deterring birds from perching but every year some – especially large birds of prey – fall victim to electrocution; their burning bodies have even been implicated in starting huge forest fires.

❓ Is bread bad for ducks – and if so, what should we give them?

I have certainly felt guilty about feeding week-old bread to pathetically grateful ducks but have convinced myself that it probably makes excellent weed sandwiches. I am therefore somewhat relieved to learn from Sonja Taylor-Jones of the Wild-fowl & Wetlands Trust at Slimbridge that stale bread isn't all that bad for ducks – though it's better if it's wholemeal. Mouldy bread is best avoided, as the spores can cause respiratory infections.

The main problem with bread, it turns out, is that everybody dishes out the stuff, leading to an unbalanced diet and large quantities of bread floating uneaten on the pond which, as it breaks down, can promote the growth of algae and cut the oxygen content of the pond to unhealthy levels. It is, therefore, a good idea to leave the bits of bread by the side of the pond, rather than in it.

At the W&WT they give their wildfowl a mixture of wheat, barley, millet, canary seed and a specially formulated pellet made up of grass, fish meal and cereal. They therefore recommend taking a bag of mixed grain along to the pond rather than the standard old loaf of bread, to give the ducks a bit of variety.

❓ If evolution is random, how can it account for the design seen in nature?

The late, great cosmologist Professor Sir Fred Hoyle raised a similar point in the 1970s, saying there was just not enough time for randomness alone to create such wonders as human beings and so the universe must have existed forever. However, Sir Fred had fundamentally misunderstood Darwinian evolution. It is true that genetic mutations in living creatures are random but this is only half the story. The other half lies in natural selection, the process which weeds out all but those mutations which boost the reproductive success of the organism. It is an astonishingly powerful filter, leading to the rapid emergence of creatures that seem uncannily well-suited to their environment. Even so, there is no guiding mechanism behind it all (or, more accurately, no absolute need to invoke one), other than the certainty that duff mutations get weeded out in the long run. As so often with evolution theory, all this sounds like a *Just So* story, so let me add that the same combination of random mutation plus natural selection is now routinely used by computer scientists to "evolve" astonishingly impressive solutions to a wide range of problems, from airline schedules to new product designs.

❓ How do cats survive falls that would kill humans?

There are many stories of cats surviving falls from great heights with only relatively minor injuries. Veterinary scientists refer to it as "High Rise Syndrome" and have written learned papers on the subject, describing incidents in which cats have fallen from thirty-two storeys up and got away with nothing worse than some chest injuries and a chipped tooth. These studies seem to prove that the vast majority of cats do survive such falls but there is an obvious selection bias at work: to wit, vets tend not to see those moggies who were killed. Still, of those cats that survive long enough to see a vet, most do indeed get to live another day. The statistics of the injuries sustained cast some light on the secret of cats' success. Part of it lies in their well-known ability to twist in mid-air and arrange a feet-first landing. For falls below around six storeys, cats have a tendency to brace themselves for impact, resulting in broken legs. Above that level, they seem to relax and allow their legs to spread out,

125

increasing their aerodynamic drag. Having turned themselves into furry parachutes, they then descend at a constant speed of around sixty-five mph. As such, it makes little odds whether the cats fall from six storeys or twenty-six: they will hit the ground with the same amount of bone-crunching kinetic energy and chances of survival. Humans, in contrast, cannot pull off such aerobatics, giving them a terminal velocity around twice as high – and thus an impact energy four times that of cats. The great biologist J B S Haldane put the situation succinctly, if graphically, in his 1928 essay *On Being the Right Size*: "You can drop a mouse down a thousand yard mine shaft and on arriving at the bottom it gets a slight shock and walks away. A rat would probably be killed, though it can fall safely from the eleventh storey of a building; a man is killed, a horse splashes". Knowing Haldane, he probably did the experiments to prove it.

? Why does a duck's quack not have an echo?

This one really baffled me – not so much in trying to find an explanation but in wondering why anyone ever believed it to be true. Having walked round a good few duck ponds with my dog over the years, I feel sure I have heard the echo of quacks from the underside of bridges and the like. However, a quick trawl of the web suggests that there is a widespread belief in the echo-less quack, so I had better try to answer it. One obvious point to make is that the existence of a noise is merely a necessary but not sufficient condition for the existence of an echo: it must be emitted in a place where an echo can be heard. Ducks traditionally spend their time on flat ponds surrounded by soft, sound-absorbent, material like bushes and trees – not exactly prime echo-generating conditions. Another possible factor was identified by some scientists at Salford University recently: the staccato pattern of the noise and the way it tails off at the end tend to mask any faint echoes that might exist. Enough already: the quack of a duck does not constitute a challenge to the tenets of physics.

? Is it true that moss tends to grow on the north side of trees?

According to many hardy outdoor types, it's possible to use moss on trees to work out which direction is north. The trouble is, these survival experts don't agree on which side of the tree the moss should be. Many claim it is the north side, arguing that moss prefers the cooler, darker, wetter conditions found there in the northern hemisphere but some insist moss grows thickest on the south, sun-facing side. Worse still, it seems entirely possible for local conditions to produce suitable conditions for moss growth pretty much anywhere on the tree. As such, I for one would be reluctant to put my trust in this bit of folklore.

? How do sea mammals like dolphins survive only on salt water?

For humans, drinking sea water has the paradoxical effect of causing death through dehydration. As the salt content of bodily fluids increases, the fresh water in living cells starts to leach away in an attempt to balance the salt concentrations on either side of the cell membrane (a process known as osmosis). This prompts a demand for more fresh water: a vicious circle that begins with terrible thirst and ends in death. How sea mammals avoid this is a mystery. Analysis of the bodily fluid of sea mammals like dolphins shows that it is pretty similar to that of land mammals with ready access to fresh water, including humans. This suggests that they have found some way around the problem of being immersed in a huge amount of a fluid they cannot drink. One possibility is that they obtain all their fresh water via their food. Certainly, sea-lions fed on nothing but fish – whose bodily fluid has a relatively low salt content – seem to get all the fresh water they need. Another possibility is that sea mammals have evolved some way of ridding themselves of the salt in any water they do take in. Again, studies of seals tend to support this, as their urine is twice as salty as seawater. Even so, for most sea mammals, scientists have little clue how they turn brackish water into Adam's wine.

❓ How does moisture from the soil reach the top of tall trees?

I thought everyone knew the answer to this: capillary action, in which the surface tension of water drives moisture up the many narrow tubular vessels within the tree. There is a problem with this explanation: were it true, the world's tallest trees would be the size of a pencil. Putting in the numbers, capillary action is likely to raise water in trees by just a few inches at best. Whatever the means by which the 350 ft-plus redwood trees of northern California get their moisture, capillarity isn't it.

The real answer is transpiration, in which the water inside trees is pulled upwards by the effect of the sun's heat. The myriad columns of water deliver nutrients and moisture throughout the tree until they reach the leaves, where the water evaporates into the surrounding air. So powerful is the effect that it allows water to defy both gravity and the frictional drag of the plant tissues. Inevitably, however, there does come a point where the process simply can't hoist the water any higher – not least because the column of water is torn apart under the competing forces. Even so, a study published in 2004 in the journal *Nature* showed that transpiration could still allow even the world's tallest trees to grow another 50 ft, to a giddying height of 425 ft.

❓ Are all animals usually right-handed?

Around one in ten humans are left-handed and studies of depictions of humans spanning over 5,000 years suggest this proportion has essentially remained unchanged throughout recorded history. Such a strong propensity towards right-handedness is, however, unique to humans. While many individual animals do show a preference for using one hand (or, rather, paw), the relative proportions are typically around 50:50. Our closest evolutionary relatives, chimpanzees, are ambidextrous in the wild. While they do show a slight preference for using their right hand when performing high-dexterity tasks, the proportions are just 60:40, compared to the whopping 90:10 figure for humans.

It is tempting to think that perhaps humans are just the end result of an imbalance that had its origins in the slight right-handed bias of chimps. However, Professor Chris McManus of University College London, the author of the brilliant (and award-winning) *Right Hand, Left Hand* (Weidenfeld and Nicolson, 2002), puts paid to that idea, by showing that much more primitive animals such as mice also show the same slight bias towards right-handedness. Quite why it exists remains a mystery.

? Why can't we tickle ourselves?

Charles Darwin argued that our wriggling and squirming response to tickling is part of a natural reflex that helps us escape from attackers when they've grabbed vulnerable parts of our body. As such, it is impossible to tickle ourselves because we know what we're going to do and where, which hardly constitutes a threat. In 1998, Darwin's arguments were confirmed by Dr Sarah-Jayne Blakemore at the Institute of Neurology, London, who found that information on exactly where we are about to tickle ourselves is sent from the cerebellum to the somatosensory cortex (the part of the brain which deals with touch sensations). Having been tipped off in advance, the brain doesn't register any threat and so we don't end up as giggling wrecks.

❓ Why are women smaller than men?

In the UK, the average height of adult women is 5ft 4in, around 5 inches shorter than the average for men, yet, as gangly supermodels show, there is no law of physics preventing women from being taller than men. However, there may be some biological and evolutionary influences at work. First, girls experience a shorter period during which their long bones can grow. Second, there are obvious reasons why women may have acquired a preference for taller men, who would have made better hunters.

As brains are now more important than brawn, this selection effect appears to be diminishing: a recent study by Australian academics showed that the average height of women has increased by one per cent over the last eighty years.

❓ What is the meaning of "double-jointed"?

Having discovered the ability to bend the upper half of my thumb until it makes a right-angle, as a child I felt rather chuffed to think I had two joints where most people had one. The correct medical term – hypermobility – is no less exciting, though it simply means that one possesses highly elastic ligaments which allow the joints to reach angles beyond most people's. Those with hypermobility can perform all sorts of amazing feats, such as bending their thumb right back until it touches their forearm and their knees left and right as well as forwards

Many children have some hypermobile joints, because their ligaments have yet to grow strong enough to restrict movement (which explains why eight year olds can sit in the lotus position with impunity) but everyone can attain some level of hypermobility through exercise, especially yoga. It is a trait that tends to run in families, especially in those with tall fathers; scientists suspect there is a strong genetic link. There is a cost to hypermobility however: those with the syndrome often suffer more than their share of injuries, aches and pains and run a higher risk of developing osteoarthritis.

? Could huge dinosaurs really sustain themselves on a veggie diet?

Short of finding the supermarket bill of a brachiosaur, there might seem little hope of working out precisely how a 75ft long, 50 ton monster fed itself but it is possible to make a few guesstimates, by assuming the present laws of physics and chemistry applied sixty-five million years ago. Dinosaurs can be regarded as being pretty similar to us: thermodynamic engines propelled by food energy.

For a very rough estimate, we simply divide the mass of the brachiosaur by that of a modern creature like an elephant and assume the amount of food consumed is pro rata. Elephants eat around 350lbs of vegetation a day, so with a mass sixteen times greater, a brachiosaur would have needed to munch its way through up to 2.5 ton of green stuff. In reality, the figure is likely to be less than half that. First, detailed studies on modern animals suggest that their energy demands are proportional to their mass raised to the power of 0.75 ($m^{0.75}$), so the brachiosaur will need only eight times as much food as the elephant. Second and much more important, brachiosaurs may have had a more lizard-like metabolism, which burns up energy at less than a tenth of the rate of warm-blooded mammals. Thus, despite its colossal size, it's possible the brachiosaur could have got away with munching through a similar amount of food to an elephant and perhaps even less, if there was a form of vegetation back then which crammed in far more energy per pound than today's varieties.

? Why don't spiders stick to their webs?

Spiders produce a range of threads when constructing their webs, only one of which is sticky. The glue lies along the thread in globules, between which the spider daintily steps as it heads towards its prey.

? Why can't you train cats?

Cat owners would say that cats are too smart to perform stupid tricks like a dog. According to Colin Tennant, one of Britain's leading cat experts, the real explanation lies in evolutionary history. Dogs were originally pack animals, who learned to pick up on cues from those around them, while cats were solitary hunters who never needed to acquire this ability. Cats will respond to the sound of food tins being tapped but few will deign to do much else.

? How do you tell stinging nettles from the harmless variety?

Genuine nettles all sting but the so-called "dead nettle" – an altogether different and harmless plant – looks very similar to the stinging variety and often grows among them. According to Derek Hilton-Brown of the wildlife charity Cone, dead nettles typically grow to about 60 cm in height and have pretty white flowers surrounding their menacing-looking leaves. Stinging nettles, in contrast, grow much taller – and just look menacing.

As with so many of nature's poisons, the precise mix of chemicals responsible for the sting of a nettle is unknown. It was once believed to be primarily formic acid (the same as that found in ants) but this appears to be a relatively minor component compared to the histamines, acetylcholine, serotonin and an unknown fourth compound so far identified. The result is familiar enough: a burning, itchy, semi-allergic response that gives the nettle its generic name, *Urtica*, from the Latin meaning "to burn". Being stung by British nettles is little more than mildly irritating but the varieties that grow in the Far East cause symptoms similar to lockjaw that last for many days and in some cases prove fatal.

? Why are eggs egg-shaped?

There's a widespread belief that the shape is the result of the egg-laying process but this fails to explain why some birds, such as ostriches, produce perfectly round eggs. A similar criticism can be made of the idea that the shape is dictated by the need for eggs to be as strong as possible: if this were the case, all eggs would be spherical but pigeons and diving birds lay eggs that are pointy at both ends.

The shape of living things is often a reflection of a sophisticated evolutionary effect and recent research suggests the same is true of eggs. Dr Tamas Szekely, a mathematician at Bristol University, has shown that the explanation appears to lie in the *numbers* of eggs laid by different bird species. Specifically, he found that eggs take on the shape they do in order that as many as possible can be packed together for warmth during incubation.

For example, shorebirds like plovers typically lay approximately four rather pointy-shaped eggs which, according to Dr Szekely allows the eggs to be around eight per cent larger for a given nest size. In contrast, birds like ostriches, which lay one egg, do best by having spherical eggs. Inevitably, there are exceptions to the "optimal packing" rule. They include guillemots, whose single eggs are pear-shaped. This may be a reflection of the fact that guillemots make their nests on sheer cliff faces: pear-shaped eggs don't roll in a straight line, thus helping to stop them rolling over the edge.

? How does a spider manage to spin a lengthy line overnight across a wide gap?

I have often puzzled about this as I blunder into a line of thread with no obvious means of support while making my way to the shed at the top of the garden early in the morning. It is a trick perfected by around 3,000 species of spider and, as with so much else about spiders, the details can be found in Paul Hillyard's delightful *Book of The Spider* (HarperCollins, 1994). The answer is, well, blowing in the wind: the spider climbs up to an exposed spot – like the top of a fence – works out where the breeze is coming from and feeds out an extremely fine thread. Like an angler, the spider then waits until it can feel some tension in the thread, showing that it's caught on to something on the other side of the gap. It then pulls this "bridge thread" tight and scampers across several times, laying down extra threads, which rapidly increase its thickness and strength. From the midpoint, the spider then drops down, releasing a second thread on the way, which it fixes at a lower level. Pulling this tight creates a Y-shaped structure, around which it can start building its web. Some species of tropical spider make webs over two metres across. I suspect, however, that many of those long lines of thread we find across our path are the result of a spider being unlucky with the breeze and only finding a point of attachment at some ludicrous distance from its starting point.

? How do autumn leaves produce such spectacular colours?

For many years, textbooks explained the colour of autumn leaves in terms of the death and decay one might expect with the approach of winter. With summer gone, the sunlight-trapping green pigment chlorophyll begins to break down, allowing others – such as the carotenoids, responsible for the spectacular yellows and golds of many autumn leaves – to reveal themselves.

Recent studies have shown that this theory is a bit glib and Professor David Lee of Florida International University and Professor Kevin Gould of the University of Auckland have put forward new research into the red coloration of autumnal leaves.

For many trees, this is due to another pigment, anthocyanin. Its appearance was also thought to be due to the decay of cholorophyll but now it seems to be part of the tree's efforts to cope with the changing weather. During autumn, the combination of cold and bright weather starts to affect photosynthesis, by which trees turn carbon dioxide and water into nutrients. This can cause permanent damage to the leaf cells and premature death. Research published in 2002 by Lee and others showed that the red anthocyanins act as a sunscreen, absorbing the excess light energy and stopping chlorophyll molecules from being torn apart. Anthocyanins also seem to be good at mopping up damaging free radicals created in the plant during cold, bright days.

This new research raises a new mystery: why did trees evolve such a sophisticated way of protecting leaves that are about to drop off anyway? Preliminary studies by Lee and Gould suggest that anthocyanins have another role in helping the tree to move nitrogen from its leaves and into the branches and trunk. What is clear is that the spectacle of autumn leaves is more sophisticated than many scientists once thought.

? Is it true that some chimpanzees have been taught to "speak"?

I can still recall being mesmerized by work in the early 1970s which described how a chimp named Washoe had been taught American Sign Language (ASL) and was using it to converse with her trainers. Since then, scientists have claimed to have taught ASL to many great apes and to have seen clear signs of intelligence in the way the chimps use words – for example, inventing the combination "cry hurt food" to describe the nasty taste of radishes. Washoe mastered over 200 signs and even taught ASL to her adopted son without human intervention. The implication is that the only reason chimps don't chat with us about the state of the economy and the awful weather is that they lack a sophisticated voice box. At least, that is what some of the researchers involved in these projects want us to believe. Many other scientists insist that these claims really tell us more about the wishful thinking of researchers. Critics of the project dismissed the idea that the chimps had learned ASL, arguing that many of the signs were either meaningless or just standard chimp gestures given an ASL interpretation by over-enthusiastic observers. Linguists insisted that there was very little evidence that the chimps were using ASL to form anything but the simplest phrases, like "Banana me me me eat", despite the fact that ASL is a full-blown language with grammar and syntax. While some of the claims for the abilities of Washoe and her chums are overblown, some of the criticisms seem tailored to ensure language remains a uniquely human trait. Perhaps the fairest summary is to say that some chimps can express simple thoughts using sign language but there is no good evidence they are sparkling conversationalists.

? How can whales dive to great depths without **•** apparent ill-effect?

The diving abilities of some air-breathing mammals is truly breathtaking: seals, dolphins and porpoises are all capable of diving to over 1,000 m. Sperm whales have been recorded at depths of almost 1.5 miles and the stomach contents of one sperm whale suggests it had been feeding off a type of dogfish only found at depths of around 1.9 miles. At such depths, pressures reach around 300 kg per square cm, so whales are performing something miraculous down there to come back up alive. No one knows for sure how they do it but studies on elephant seals suggest that deep-diving mammals don't bother to resist the pressure and let their lungs collapse as the pressure builds. This may hold the key to their ability to avoid "the bends", caused when nitrogen gas crammed into the bloodstream by the enormous pressure starts to form bubbles in tissue on the way back up. As their lungs collapse, the nitrogen-rich air the animals breathed in at the surface is forced into their windpipe and nasal cavities, out of harm's way, and is only allowed back in very gradually during the ascent. Even so, sea mammals do run the risk of getting the bends if they ascend too rapidly. In December 2004, scientists from the Woods Hole Oceanographic Institution in Massachusetts announced that a study of bones from a collection of sperm whales suggested the creatures had endured mild but chronic decompression sickness over the course of their lives.

❓ How do cats purr?

Vets used to think that the noise was caused by vibrating blood vessels at the back of the cat's throat. A slightly more plausible explanation links it to rapid twitching in muscles attached to the cat's voicebox. As the cat breathes out, air passes through its vibrating voicebox, producing the oddly comforting sound we call purring. The frequency of this sound may at least explain *why* cats purr. For a house moggy, it's between 27 and 44 cycles per second – strikingly similar to the type of vibration doctors have found encourages bone healing in humans. This has prompted scientists to suggest that purring cats may be repairing themselves after a hard day's mousing. If true, then I have a suggestion: make cats available on prescription, so they can sit on the laps of the elderly and prevent hip fractures and osteoporosis.

❓ Why do some people sneeze when they look at the sun?

This curious reflex action, sometimes known as *photic sneezing*, affects around 1 in 4 people – including me. It may be inherited but as far as anyone can tell, it's completely useless – just like those other bizarre inherited traits like the ability to wiggle your ears or curl your tongue. The culprit seems to be the trigeminal nerve, which plays a key role in many involuntary actions, including sneezing. This nerve has many branches, most of which end in the skin on the skull. When stimulated by, say, a speck of dust near the nose, this nerve sends signals to the brain, triggering a sneeze. Some of the branches also extend up to the eyes, where they respond to irritants, such as onion vapour or ammonia, by producing tears. In some people – for reasons no one quite understands – these branches also respond to bright light, resulting in a sneeze. Whatever the cause, a bout of photic sneezing can certainly be annoying and possibly dangerous, especially if it strikes while driving. Like the coughing reflex, with a bit of determination it is possible to prevent the sneezes going on too long.

❓ Whatever happened to the claims for "gay genes" a few years ago?

They did what so many of these genetic stories do – made a lot of headlines, then fizzled out. In 1993, Dr Dean Hamer of the US National Cancer Institute created a sensation by uncovering evidence for the existence of genes linked to homosexuality in pairs of gay brothers. Around three-quarters of the brothers had the same set of sequences on part of the bundle of DNA in their cells, known as the X-chromosome, suggesting that at least one gay gene existed within this region of their genome. Two years later, Dr Hamer claimed to have replicated the findings. Other studies also found that identical twins had a much higher rate of producing two gay boys than non-identical twins, which is consistent with the idea that there is a genetic influence. However, in 1999 the largest-ever gene-based study of twins by researchers at the University of Western Ontario, Canada, failed to replicate Hamer's original findings, leaving everything up in the air, which is where it remains to this day.

❓ Why did humans in hot countries evolve dark skin?

There's a lot more to sunlight than heat, and evolution seems to have focused on these other qualities rather than fret about heat, which we can control relatively easily anyway by staying in shade on really hot days. While there is still much debate over the origins of dark- and light-skinned humans, it seems clear that the answer lies in the properties of melanin, the chemical responsible for skin colour. Melanin is known to be protective against ultraviolet light, whose cancer-causing abilities are strongest in hot, low-latitude countries. So, having melanin-rich skin makes a lot of sense near the Equator. On the other hand, ultraviolet light is needed to produce Vitamin D, deficiency of which can cause the crippling bone disease rickets. Having high-melanin skin in higher latitudes thus increases the risk of acquiring this disease (as was demonstrated early last century by the ninety per cent prevalence of rickets among black infants in New York). Natural selection thus has to achieve a balance in optimizing melanin levels. To judge by the distribution of light- and dark-skinned populations, it seems the principal sunlight-related challenge in equatorial countries is cancer, while in more northern latitudes, it is lack of Vitamin D. Modern science has greatly reduced the importance of these evolutionary pressures: UV-blocking sunscreens allow even the palest Scandinavians to live in Equatorial countries, while the addition of Vitamin D to milk has made rickets virtually unknown in northern latitudes.

Chapter 7

The Earth below, the sky above

❓ How did the Ancient Greeks know the Earth was round?

It's still possible to find history books which blithely talk of how Columbus first showed that the Earth was round but as early as the fourth century BCE, Aristotle was declaring that "The sphericity of the Earth is proved by the evidence of our senses". He put forward several different ways of showing that we live on a giant ball. For example, explorers reported new patterns of stars coming into view in the night sky and familiar ones vanishing over the horizon as they travelled further from home, no matter which way they went. Even more tellingly, as a ship approached its home port, those waiting for it would not see just a tiny ship in the distance getting ever larger but would first see the tip of its mast and then ever more of its hull – just as if it were sailing over the crest of a globe. A century later, Eratosthenes succeeded in measuring the circumference of the Earth. By observing the length of the sun's shadow from two different cities on the same day of the year, he found a figure of around 28,500 miles – within fifteen per cent of the true value of 25,000 miles. Columbus, in contrast, based his expedition plans on the far smaller figure of 18,500 miles – probably as a ruse to boost his chances of sponsorship.

? How will continental drift affect future maps of the Earth?

The idea of continents meandering across the face of the Earth is surely one of the most mind-boggling in all science and it is hard not to feel sympathy with the distinguished academics who dismissed it when it was mooted by the meteorologist Alfred Wegener a century ago. Nowadays, even school science texts include maps showing the changing face of our world over the last few hundred million years. Continental drift is still underway, propelled by the roiling currents of magma beneath our feet. According to satellite measurements, Europe and America are separating at the rate of a few centimetres a year (about the rate at which fingernails grow).

Given the different directions in which the various continents are headed, projecting what the world will look like in the future is no easy task but Professor Christopher Scotese and his colleagues at the University of Texas, Arlington have had a shot at it, with fascinating results.

Approximately fifty million years from now, the most striking change to have taken place will be the disappearance of the Mediterranean Sea, as Africa will have smashed into Europe, closing up the sea-filled gap. America and Europe will have grown much further apart but Eurosceptics will be delighted to know the UK will still not be part of Europe: indeed, according to Scotese's calculations, Britain will still be refusing to join Europe even 250 million years hence, when all the rest of the world's continents will have fused together to form a "supercontinent".

? When will fossil fuels run out?

One lunch hour in the early 1980s, I had what could euphemistically be called a frank exchange of views with a colleague over the claims being made by an the economist, Julian Simon. My colleague was extolling Simon's thesis, set out in his book *The Ultimate Resource* that, contrary to doom-laden assertions, the world will never run out of natural resources. According to Simon, human ingenuity has always ensured that long before any resource literally runs out, some alternative is found. This seemed like baloney to me but then I learned of a bet Simon had made with the famously doomy demographer Paul Ehrlich. In 1980, Simon had wagered that the price of any five industrial metals chosen by Ehrlich would actually be lower in real terms ten years later – in line with his contrarian view that nothing ever really runs out. Ehrlich accepted what seemed an easy way to relieve smart alec Simon of money. Then 1990 arrived and the prices of the five metals had indeed fallen by an average of almost forty per cent in real terms compared to 1980.

Simon's successful prediction has made me very sceptical about claims that everything is running out (it had a similar effect, I gather, on Bjorn Lomborg, author of *The Skeptical Environmentalist*). While researching the answer to this question, I was thus not remotely surprised to learn that fossil fuels have also been predicted to run out many times over the last century, while doing no such thing.

As early as 1874, there were warnings that US oil reserves would dry up within four years. In 1920, the US Geological Survey estimated the world's total reserves of oil at around sixty billion barrels; barely two years' worth at today's rate of consumption. The figure zoomed upwards to 600 billion barrels in 1950 and today it is five times higher still. During the 1940s, 35 per cent of drilled wells were dry. By the 1990s, just 23 per cent were. Long term, the price graph has seen pretty steady – except for occasional blips like 1973 and 2005.

It is a similar story with coal and gas, the price of which is declining in real terms. How is this possible? According to a report issued in 2004 by the US-based National Centre for Policy Analysis, the explanation lies in the ever more sophisticated methods of finding and extracting the stuff, just as the late Professor Simon claimed. Is this an excuse for profligacy? Not at all; quite apart from the environmental arguments against waste, there are some geopolitical ones. For example, if the US could give up its gas-guzzling ways and become as energy efficient as western Europe, it could free itself of reliance on oil from the Gulf states.

? Where did all the water in the world's oceans originally come from?

Explanations for how 320 billion billion gallons of water ended up on the Earth's surface after its formation have undergone something of a, well, sea change since the 1970s. It is generally thought that much of the water came from ice trapped inside dust grains that made up the primordial cloud of material from which the Earth condensed around 4.5 billion years ago. Until relatively recently, it was thought that this trapped water was released by volcanic eruptions, which led to huge clouds of water vapour forming around the planet. These cooled and released their content as rain, which over hundreds of millions of years filled up the basins which now form the oceans.

In his outstanding account of modern oceanography *Mapping The Deep* (Sort Of Books, 2000), Robert Kunzig says that research now suggests that such volcanic outgassing could not release sufficient water vapour to form the world's oceans. Instead, the primordial ice is thought to have been turned into water during the catastrophic impacts that took place during the formation of the Earth. Comets are also believed to have made a contribution, dumping the ice they contained on to the Earth. Indeed, perhaps as much as half of the water on our planets came from comets – something to ponder the next time you fill the kettle.

? Will the Earth ever stop spinning?

Left to its own devices, the huge mass of the Earth would probably cheerfully spin forever in the frictionless vacuum of space. As it is, the Sun and Moon are steadily slowing it down through the tidal drag they exert on the world's oceans. The effect is very small: a day lasts barely 0.0017 seconds longer than it did a century ago – negligible on human timescales. It does, however, add up. For example, when dinosaurs ruled the world, the planet was spinning markedly faster than it does today, making a day around half an hour or so shorter. Studies of the daily growth rings on fossilised coral and molluscs have also shown that the day was just 22 hours long around 450 million years ago.

At the current rate, it will take billions of years for the Earth finally to come to a halt, by which time humans will probably have left for somewhere better. This assumes, of course, that the rate of spin will not change, which is by no means certain. Indeed, it lies at the heart of a puzzle about the past history of the Moon. As the Earth spins ever more slowly, it loses angular momentum to the Moon, which moves slightly further out in its orbit. The predicted increase is very small but it has been detected by bouncing laser beams off equipment placed on the Moon by the Apollo astronauts, revealing that our nearest celestial neighbour is currently receding from us by around 1.5 inches per year. Again, this is negligible on a human scale but it tots up – with perplexing results. Calculating the effect of the Earth's spin-down on the Moon's orbit suggests that the two must have been in contact around 2,000 million years ago. Astronomers are fairly confident, however, that the Moon has been a separate entity for at least 4,500 million years. Clearly something has gone wrong with the theory somewhere but quite what is far from clear.

❓ How long has Antarctica been frozen over?

Almost a century ago, the expeditions of both Robert Scott and Ernest Shackleton found evidence, in the form of fossil ferns and coal, that Antartica has not always been a frozen wilderness. Even more impressive evidence emerged in 1991, when scientists uncovered the remains of a dinosaur, *Cryolophosaurus* which roamed the continent around 190 million years ago. The date ties in with geological evidence that Antarctica was then basking in equatorial heat, being part of the huge "supercontinent", Gondwana. This began breaking apart around 160 million years ago and what is now Antarctica, Australia and South America drifted south. After around 100 million years, they had all reached their current locations. It took another forty million years for Antarctica to become the coldest place on Earth, by which time it had broken away from its neighbours and become surrounded by cold ocean currents. For the last twenty million years it has been almost permanently frozen over – there is evidence that the continent became ice-free as recently as three million years ago, though no one knows why. Perhaps it was the result of global warming triggered by all those stone age power stations.

? How did Scott and Amundsen find the South Pole without modern navigational aids?

In the case of Scott, the short answer is: he didn't. Navigation is actually relatively simple near the poles, as both Scott and Amundsen learned from A R Hinks, a Cambridge University surveying expert, who gave a seminar on the subject at the Royal Geographical Society in November 1909. As Hinks pointed out, the "bunching" of the lines of longitude at the poles means that one degree of longitude represents relatively few miles and so can be ignored. The key thing was to keep heading south, which can be done via latitude measurements made with a sextant and judicious use of a compass. Ever the professional, Amundsen took Hinks's advice and used sextant readings and dead reckoning to get him close to the pole. Once there, he and his team – which included four fully-qualified navigators – took careful sextant observations that fixed the position of the true pole to within 250 m. They then tramped over it three times to make sure.

Like so much else about his expedition, Scott took a "gentleman-amateur" attitude towards the issue of navigation. He nearly didn't bother having a qualified navigator but at the last moment invited Petty Officer Henry Bowers to join the team. Unaware of Hinks's advice, Bowers spent much of the polar journey making theodolite measurements and pointlessly long and complex calculations in atrocious circumstances. Needless to say, when they came to make the critical calculations to fix the precise position of the South Pole, Bowers and Scott made a mistake. So it is not quite true that Scott and his team arrived at the pole a month after Amundsen: they never actually reached it at all.

? Will deforestation eventually cause us to run out of oxygen?

Environmentalists are fond of calling the world's forests in general and the Amazon in particular, "the lungs of our planet", the implication being that unless we stop deforestation, we're all going to be gasping for breath. In 2002, the environment minister of Colombia went so far as to plead with the world's cocaine users to give up their habit, because they were promoting damage to the rainforests "that the world needs for its oxygen." Fortunately (as so often with eco-scare stories), this is the purest bunkum. By far the biggest contributor is the horrible, slimy, thoroughly unphotogenic algae floating like green scum on the world's oceans, which accounts for about ninety per cent of oxygen production. There are many reasons for being dead against rampant deforestation (or cocaine use, come to that) but the nightmare scenario of a suffocating planet isn't one of them.

? How much water is there in the atmosphere compared to the oceans?

The amount of water in the air obviously varies enormously according to when and where one makes the measurements but a reasonable average figure would be around three parts of water per thousand of air by mass. As the mass of the atmosphere is around five million billion tons, this means there is around fifteen million million tons of water in the air – which sounds a lot, until one learns that there is around 100,000 times more water in the oceans.

? Why will global warming raise the world's sea levels?

This does seem a strange claim, for, as everyone knows, ice takes up more volume than the equivalent mass of water (which is why it floats). So, you might expect the much-heralded melting of the floating Arctic ice cap slightly to lower the level of the world's oceans. In fact, it won't, because not all of the extra volume taken up by ice is actually immersed: some of it sits above the water-line. When the ice melts, the density difference vanishes and the resulting meltwater takes up the volume of seawater originally displaced. Result: no change in sea level.

That's not the case for the snows of Antarctica and the glaciers around the world, however, as they are currently sitting on top of dry land and thus not yet contributing to the level of the sea. In any case, the single most important effect is the thermal expansion of sea-water; combined with the other contributions, this is expected to produce an average sea level rise of 10 cm by 2030.

? Does Britain have small earthquakes that go unreported?

Despite having a reputation for being as solid as, well, a rock, many parts of Britain experience very feeble tremors (so feeble, indeed, that they have negative scores on the Richter scale of earthquake intensity). The causes of these tremors have been investigated for over a century. Some are the result of human activity, such as mining or quarrying. Then there are "microseisms", which are thought to have two principal causes. The first is the pounding of a steep coastline by the sea, which produces tremors with a typical frequency of around four per minute – characteristic of the rate at which surf breaks on the coast. The other is a more complex manifestation of coastal waves, leading to rhythmic changes in pressure on the sea-bed, which responds in a tremor-like fashion.

? Could global warming cause the Gulf Stream to vanish?

The concern is that as global warming takes hold and the Arctic ice melts, the resulting fresh water – which is less dense than the rest of the sea – causes disruption of the circulation of the warm Gulf Stream. This could lead to temperatures in the UK dropping to levels not seen since the last time this occurred, around 17,500 years ago.

In 1999, researchers at the Scottish Executive's Marine Laboratory, Aberdeen, analysed the salinity of sea water from between Shetland and the Faroe Islands collected since 1893; the results suggest a more pronounced drop in salinity in the last twenty years than in the previous century. This is consistent with an increase in the level of fresh water entering the sea from the Arctic, which appears to have reduced the strength of the warm current by around twenty per cent over the last fifty years. Yet, as always with climate studies, it's hard to tell whether this is a blip or part of a genuine trend. If it proves to be the latter, some scientists claim the effects will be felt rather sooner than one might expect. Computer models and studies of ice-cores suggest that ocean currents can switch on and off surprisingly rapidly, causing drastic temperature changes over just a decade or so – a blink of an eye in geological terms.

? The Earth is furthest from the sun in July, so why is this the hottest month of the year in the northern hemisphere?

It's certainly true that the Earth is furthest from the sun around July – some three million miles further away than in

January – and this does mean that it gets around six per cent less heat from the sun. It's clear that the Earth's distance from the sun cannot be the reason why the hottest month is when it is: if it were, then everywhere on the planet would have summer at the same time. It's principally due to the fact that our planet is tilted some 23° from the vertical. During the Northern hemisphere summer, it is tilted towards the sun, whose rays strike at a relatively steep angle, so their heating effect per square yard is relatively high. During the Northern hemisphere winter, however, it is tilted away from the sun and its rays strike more obliquely, thus diluting the amount of heat that hits each square yard. Down Under, of course, it's the other way around. So in principle, in northern latitudes, the hottest day of the year should be the summer solstice, 23 June. In Britain, in practice, it takes a while for the sea currents to catch up with events overhead and so it's usually July before the UK starts to feel the benefit.

? Why does the sun still rise later in the morning after the shortest day of the year?

This is not the only curiosity about the sun's behaviour during the winter: the earliest sunset does not coincide with the shortest day either but takes place over a week earlier. Both these anomalies are the result of two features of our planet: first, its orbit around the sun is imperfectly circular and second, its axis of rotation isn't perfectly perpendicular to the plane of its orbit. The combined effect of these is to change the apparent speed with which the sun completes its daily journey from East to West over the course of a year.

During the winter months, in northern latitudes, the Earth's eccentric orbit brings it ever nearer to the Sun, reaching its closest point ("perihelion") around 4 January, at which point it is travelling at its fastest through space. By this time, the tilt of the Earth has waded in with its own effect, and the Sun starts to move back north again for the summer. Thus, contrary to appearances, the Sun does not march across the sky at the same pace throughout the year but speeds up and slows down according to a messy combination of these two effects.

That in turn makes the Sun a less than ideal clock; to overcome its deficiencies Victorian astronomers invented a fictitious version they could rely on, called the Mean Sun, whose speed across the sky is constant. This artificial sun became the basis of Greenwich Mean Time (GMT) – the time according to the passage of the Mean Sun over the Greenwich Meridian – and it is this which is used to quote times of sunrise and sunset. One consequence of this attempt to impose order on nature is a discrepancy between the timings based on GMT and what the real sun is doing, which manifests as the mismatch between the days on which the latest sunrise and earliest sunset occur and the shortest day.

? **Could icebergs be towed from Antarctica to bring water to drought-stricken areas?**

The Antarctic ice-sheet contains around seventy per cent of the world's fresh water and with 1,200 million people lacking easy access to drinkable water, the idea of bringing bergs from Antarctica has been resurrected several times since it was first mooted fifty years ago by Dr John Isaacs of the Scripps Oceanographic Institute in California. In the late 1970s, one of the sons of King Faisal of Saudi Arabia even set up a company to investigate the idea.

The idea isn't as barmy as it seems, at least in engineering terms. The immediate objection – that the berg would melt before it reached its destination – overlooks the unusually high amount of heat required to turn ice into water. It seems likely that only a small proportion of the berg would melt during its voyage from Antarctica. Ships powerful enough to tow a decent-sized berg, weighing 100 million tons, already exist and cables strong enough to lug the thing could be made. A much tougher problem, ironically, is melting it once it arrives – to supply a population with water at a reasonable rate would require the output of a large power station. Add in the substantial costs of distributing the water to where it's most needed – which is usually far from the coast – of towing it thousands of miles from Antarctica and the principal objection becomes clear: economics. For the time being, berg-towing remains an engineer's pipe-dream.

❓ Has an earthquake ever been successfully predicted?

In February 1975, Chinese seismologists, after detecting a swarm of tremors passing through Haicheng Province, issued a warning that a major earthquake was about to strike Manchuria. Within twenty-four hours, an earthquake measuring 7.3 on the Richter scale had occurred – by which time, according to some reports, evacuations had been carried out which saved the lives of thousands of people. As with so many claims during the Cold War, there is less to this apparent triumph of Communist science than meets the eye. It later emerged that the same team had made a similar prediction the previous year but that had proved a false alarm. A year later, the team failed to predict the Tangshan earthquake, the worst of the twentieth century, which left at least 240,000 dead.

This hasn't deterred others from claiming to have cracked the problem of predicting earthquakes using their own secret recipe. In 1977, an international team of seismologists decided that a fall in seismic activity near Oaxaca in southern Mexico meant that a massive quake was about to strike. Some months later, the area suffered a quake measuring 7.7, though whether it was linked to the earlier fall in activity was never clear. In 1988, the US Geological Survey warned that there was a thirty per cent chance of a quake occurring in the Loma Prieta region of California some time in the next thirty years. The prediction came true rather sooner than expected: on 18 October 1989, a 7.1 quake struck the region, killing 62 and causing several billion dollars of damage. Again, only the most charitable observer would count this as a useful prediction.

For sheer persistence in the face of scepticism and hostility, the prize must go to a team of seismologists at the University of Athens, who have made dozens of forecasts since 1987; their claimed success rate, of seventy per cent, is given little credence.

Some years ago, I looked into the mathematics behind earthquake predictions and the requirements of a reliable prediction method, which ideally would neither miss a genuine event or cry wolf. Fortunately, major earthquakes strike relatively infrequently; unfortunately, this means that unless it is astonishingly reliable, any prediction is virtually certain to be a false alarm. Using real-life data, I estimated that for a forecast system to be remotely useful, it would have to be at least 100 times more accurate than any weather forecast has ever been. The appallingly complex behaviour of fractured rock under stresses and strains makes this a dim prospect. It is therefore a safe bet that the timing and location of a major earthquake will never be predicted effectively – not that this will stop people from trying.

? Why doesn't the Earth's magnetic north pole lie at the North Pole?

As every Scout knows, the position of magnetic North isn't the same as true North, nor does it stay fixed. Indeed, its position hasn't coincided with true North since records began over 500 years ago and until the mid-nineteenth century it was wandering ever further away. Then, about 150 years ago, it started heading back north again, weaving its way among Canada's arctic islands. Over the last ten years its northward march has accelerated markedly to as much as twenty-five miles a year, bringing it to its present position some 600 miles from the real North Pole at latitude 80°N, 110°West.

The explanation for its wandering lies in the processes that create the Earth's magnetic field, which aren't fully understood. Roughly speaking, the field is thought to be produced by a "dynamo effect", similar to that which converts the movement of a bicycle wheel into electric power. In the Earth, the movement is of vast convective loops of molten iron within the Earth's hot outer core, 1,800 miles beneath our feet. The slow rise and fall of these loops, together with effects caused by the Earth's rotation, produces an electric current and with it a magnetic field, akin to that of a simple bar magnet, tilted somewhat to the rotation axis of our planet.

According to this theory, the reason magnetic North does not coincide with true North is because the Earth's rotation plays only a secondary role in generating the magnetic field, but explaining its precise location is quite another matter. In 1995, Gary Glatzmaier, of the Los Alamos National Laboratory, New Mexico and Paul Roberts of the University of California, Los Angeles used supercomputer simulations of the Earth's interior to show that the location of the magnetic poles is the outcome of a constant struggle between the liquid outer core, which is trying to flip the magnetic field right over and the solid inner core, which fights the effect. Every so often, the inner core loses the fight and the Earth's magnetic field undergoes a complete reversal, with magnetic North becoming South and vice versa.

? What will happen when the magnetic poles of the Earth flip?

There is no doubt that, at some point in the future, the magnetic poles of the Earth will flip. Studies of ancient lava, which seals in the orientation of the magnetic field as it cools, show that such reversals have taken place twenty times over the last five million years, most recently around 780,000 years ago. The flip takes around 5–10,000 years, during which the Earth is left with little or no magnetic field. According to some scientists, another reversal is (at least in geological terms) imminent. The Earth's magnetic field has been weakening since around the time of Christ and at current rates will vanish completely within a few thousand years. Whether this is of much concern for life is rather doubtful. It is true that potentially dangerous radiation from the Sun and outer space does get trapped by the Earth's magnetic field in the Van Allen Belts, which lie between 3,000 and 18,000 miles above our planet. It's also true that the concentration of radiation in these regions is high enough to present something of a hazard to anything or anyone that lingers within them. Yet, even if these belts were to vanish during a magnetic reversal, it's uncertain whether the radiation then able to reach the Earth would cause much trouble, as the atmosphere would still mop up virtually all of it before it reached the sur- face. The most obvious effect of the vanishing magnetic field would be spectacular aurorae, as more fast-moving particles would survive to smash into air molecules at the top of the atmosphere.

? Could a landside on the Canary Islands generate a major tsunami?

In 2000, apocalyptic stories began to circulate about the possible collapse of Cumbre Vieja, the volcano on La Palma, in the Canaries. Its western flank has separated from the rest and looks set to crash into the sea. With an estimated mass of around a mil- lion billion tons, the resulting splash should be pretty spectacu- lar. Indeed, according to some scientists, it will generate a tsunami about 3,000 ft high, which will spread out to devastate the coastlines of Africa, Europe and America.

Why isn't anything being done? First, no one knows how imminent the threat is: the landslip could take place next year or in 10,000 years. Second, apart from evacuating the tens of mil- lions living in the threatened areas, it's far from clear what can be done. Certainly there is no hope of pinning or supporting so massive a structure.

Many experts believe there is a strategy that may well prove highly effective in dealing with the threat: forgetting all about it. In 2003, a committee of experts issued a communique which it hoped would "halt the scaremongering from these unfounded reports". According to the committee, computer simulations and experiments have been carried out to assess the threat from La Palma and they show that the resulting tsunami won't pose anything like the threat claimed for it. Since the communique was issued, researchers at the Southampton Oceanographic Centre have published the results of studies of ancient landslips at La Palma and elsewhere, which suggest that they happen a little at a time, rather than all at once.

? Why is the interior of the Earth still hot after several billion years?

The answer seems obvious: the Earth is still in the process of cooling down from its originally molten state. However, a rough calculation of the cooling rate of a molten ball the size of the Earth puts paid to this idea. Our planet should have become a cold lump of rock billions of years ago, yet the molten magma spewed out by volcanoes and the stifling heat at the bottom of deep mines proves this isn't so. This suggests that something is acting as a source of heat. That something is radioactivity, generated by the traces of uranium, thorium and potassium trapped inside the Earth at its formation. As they slowly disintegrate, they release particles which smash into the surrounding material, warming it up to temperatures as high as 5,500°C at the core. At the surface, all this heat leaks out at an average rate of around 150 kilowatts per square mile. In short, we are all living and working on top of a vast, simmering nuclear reactor. Scary thought.

? If the earth's interior is molten, why does the heat only reach the surface at volcanoes?

The Earth is often portrayed as being a huge ball of molten lava contained in a twenty mile thick "skin" of rock. Yet analysis of earthquake waves passing through the Earth show that it is completely solid from about 120 miles beneath the surface down to the outer core. The enormous pressure produced by these overlying layers raises melting points to the point where rock can remain essentially solid despite being at temperatures in excess of 1,700 °C. Even so, the rock can still flow when subjected to huge forces – just as the solid metal of a horseshoe becomes malleable at red-hot temperatures and can be forcibly persuaded to take up different shapes.

The Earth's crust is pretty thin – equivalent to a layer of tissue paper pasted on to an orange – but it still takes a lot of heat from below to burn all the way through it to the surface. Magma usually reaches the surface through cracks in or between the huge tectonic plates that make up the Earth's crust. The most spectacular demonstration of this is the "Ring of Fire", the circular arrangement of volcanoes around the edge of the Pacific plate.

Magma can sometimes burn right through the crust when a "mantle plume", a huge, finger-like blob of extremely hot magma forms at the boundary between the mantle and the outer core, around 1800 miles down, then breaks away and rises up to the surface, triggering a colossal outburst of volcanic activity. A mantle plume was responsible for dumping around 250,000 cubic miles of lava into what is now the Deccan Traps in India around 65 million years ago – an event of unimaginable violence, whose climatic effects helped tip the dinosaurs into extinction. Continental drift has since moved the point where the plume reaches the surface to the island of Reunion, in the Indian Ocean.

Heat escapes from the Earth at the rate of around 44,000 trillion watts – over 100 times that generated by all the world's nuclear power stations – so many attempts have been made to exploit this source of free energy. "Geothermal" energy schemes usually involve injecting water into pipes set in hot rocks and using the resulting steam to drive turbines. There is probably much scope for such exploitation, if only governments could be persuaded to take it seriously.

❓ How much less would someone weigh at the Equator?

If you want to lose weight instantly and have a nice time doing it, take a flight to Singapore or Quito. Being near to the Equator, these cities experience almost the full centrifugal effect of the Earth's rotation, which reduces the effective force of gravity. They also benefit, indirectly, from the fact that the centrifugal effect also causes the Earth to bulge slightly at the Equator, so that they are also around 8 miles further from the centre of the Earth, cutting the strength of gravity further still. Sadly, even the combined effect doesn't add up to much: around 0.3 per cent compared to the UK. In other words, a 60 kg person would weigh around 200 g less in Singapore than London. It might be cheaper just to skip lunch.

❓ As ozone is a gas, how can a supposed "hole" appear?

Ozone is indeed a gas and serves the vital function of shielding us from the most dangerous form of ultraviolet (UV) radiation emitted by the sun. Comprised of three oxygen atoms, it's formed about ten to thirty miles up in the stratosphere by the impact of UV on oxygen molecules, some of which break apart and re-group in sets of three atoms. The resulting ozone mops up the most harmful form of the radiation, UV-C, which would otherwise cause severe damage to living cells.

In the mid-1980s, scientists at the British Antarctic Survey discovered that the concentration of ozone over the southern continent plummeted by around thirty per cent during October. Analysis showed that a large area of ozone had been destroyed, as if a giant hand had reached down and scooped it out. To that extent, it makes sense to speak of an ozone "hole", in just the same way as we talk of a hole in the ground. Subsequent measurements revealed that the hole was getting deeper and wider, with worrying implications for UV levels in the southern hemisphere. Various explanations were put forward to explain its existence, including the idea that some weird form of atmospheric circulation was removing the ozone, like a spin-drier removes water. The real cause turned out to be ice crystals high in the atmosphere over Antarctica, which gave added potency to ozone-destroying industrial chemicals, CFCs. These have now been banned and current measurements of the ozone hole suggest it may be filling back up again.

? How different would conditions on Earth have to be to rule out life?

This is a key issue in the debate about whether life can exist in other places in the cosmos, for if life is very picky about the conditions under which it will thrive, the odds on finding it elsewhere plunge. While biologists can argue forever about what exactly is required, there is a general consensus that life on Earth requires liquid water. This puts a limit on the temperature of the Earth: too hot and water will evaporate, too cold and it will turn to ice. This in turn puts a limit on the distance the Earth can be from the Sun; estimates vary but they typically lead to a "Habitability Zone" that begins with orbits around twenty per cent smaller than the Earth's and extends out to about sixty per cent larger – not a very big range at all, barely including even Mars. A sample of one is hardly evidence but of the 100 or so planets to date discovered around other stars, not one lies within the Habitability Zone of its parent star. As these stars are all on our cosmic doorstep, it's still far too early to put a reliable figure on the number of planets in our galaxy that could sustain life as we know it but the evidence is not encouraging.

? Can atmospheric effects allow one to see "over the horizon"?

The usual way of working out the distance to the horizon assumes that the Earth is a perfectly round, airless ball, in which case light travels in straight lines and a bit of geometry shows that we can see to a distance given by the following formula: take your altitude above sea-level in feet, add fifty per cent and take the square root; the result is the horizon distance in miles. By this reckoning, a six-foot person standing at sea-level can see around three miles. However, this ignores the Earth's atmosphere, which bends light rays and, in principle, could allow us to see somewhat further. As the light-bending ability of air at sea-level is barely 0.003 per cent different from that of a vacuum, it would hardly seem worth including. Yet, amazingly enough, detailed calculations show that this refraction effect allows light to reach us from distances around ten per cent further away than the simple formula suggests. Put another way, the atmosphere effectively increases our altitude by about twenty per cent as we scan the far horizon, so from the top of a 3000 ft peak, we can see up to five miles or so "over the horizon".

? Where does the salt in the sea come from and is it still accumulating?

The sea contains a host of compounds, washed into it by rain dissolving chemicals in exposed rocks. Those most likely to find their way into the sea are, naturally enough, the most water-soluble, namely chlorine and sodium – the raw ingredients for common salt; these make up almost ninety per cent of all the dissolved compounds in the sea. Over millions of years, the erosion process has dumped the equivalent of around four teaspoons-full of dissolved salt in each pint of sea water – a concentration which, if it were extracted from the world's oceans, would produce a layer of salt 150 ft thick. This might sound a lot but calculations of run-off rates suggest that by now the sea should have become totally saturated with salt – and as dead as the Dead Sea. It patently isn't, so the mystery is not so much why is the sea salty as why isn't it packed with the stuff?

One explanation was put forward during the 1970s, by the distinguished ecologist James Lovelock in developing his "Gaia Hypothesis", according to which living organisms interact with the Earth in ways that keep the planet suitable for life. Excessive salinity poses a grave threat to marine life and Lovelock wondered if some organism could be maintaining salinity in the face of global land erosion. He found a candidate in a primitive microbe responsible for creating huge, shallow lagoons in coastal regions such as Baja California. The sun's heat on these lagoons drives off the water, leaving the salt trapped. Whether the process is potent enough to maintain the saltiness of the sea is unclear but it is an intriguing solution to an otherwise baffling mystery.

? What percentage of the atmosphere did Concorde fly above?

Enthusiasts for the late, lamented supersonic aircraft emphasize its world-beating cruising speed of 1,350 mph but the height at which it flew was no less astonishing: 60,000 ft, almost twice the cruising altitude of conventional passenger aircraft; high enough to see the curvature of the Earth. This put Concorde well into the stratosphere, above approximately eighty-five per cent of the atmosphere, where the density of air is barely one-third of that available to conventional jets.

Chapter 8

Heavens above

? Where does space start?

More than forty years after astronauts started exploring space, there's still no internationally recognized definition of where they've all been. NASA has long had a tradition of awarding anyone who reaches an altitude of fifty miles "astronaut's wings", certifying that they have been into space. During the 1960s, eight X-15 experimental rocket-plane pilots joined the astronauts of the Mercury, Gemini and Apollo programmes in being awarded this accolade, with pilot Joe Walker reaching heights of over 62 miles on two flights in 1963. Most space experts agree missions to this latter altitude, equivalent to 100 km, constitute genuine spaceflight and it may yet became the legal standard: in 2002 Australia became the first nation to adopt 100 km as the point where space begins.

? Why do we not feel the Earth's motion as we spin round the sun?

It may be hard to credit but we're all zooming around the Sun at over 66,000 mph with no visible means of support. The reason we aren't aware of it is that it is not velocity that gives us the sensation of movement but *changes* in velocity – that is, acceleration. It's true that as we orbit the Sun, our speed is constant but our direction continually changes, producing an acceleration but the radius of our orbit is, at 93 million miles, so vast that, even travelling at that huge speed, the acceleration is 1,700 times weaker than that due to the gravity holding us on to the Earth and so far too feeble for us to notice.

? Why does the Moon seem bigger when it is on the ● horizon?

Often called the Moon Illusion, this has puzzled people at least since the time of Aristotle. Many people suspect it's the result of refraction when the moon's light comes through the thicker part of the atmosphere but while refraction can have some effect, it tends to be chiefly parallel to the horizon, producing the squashing effect familiar in clichéd pictures of the setting sun. If you actually measure the size of the moon when on the horizon and overhead, it is more or less the same, demonstrating that it's an illusion we're witnessing, not reality. The most commonly cited explanation centres on an optical effect first pointed out by Mario Ponzo, a perceptual psychologist, in 1913. When we look into the distance, parallel lines such as railway tracks and roads seem to converge at the "vanishing point" as they disappear and so we tend to think of objects close to the vanishing point as being far away. In addition, distant objects look smaller than normal; according to Ponzo, this can cause problems when we see something that we *think* is far away but that hasn't changed in size. This is just what happens with the moon: its size and distance from us are pretty constant but when it appears close to a vanishing point the brain decides it must be far away and so is forced to conclude that the reason it still takes up the same amount of sky is because it has expanded. That said, it's unclear how the Ponzo effect explains why pilots above the clouds, unable to see any vanishing point, still see the Moon as looking bigger when it is close to their horizon. This makes me suspect that the Moon Illusion arises because, in everyday experience, seeing an object on the horizon tells us it is further away than if it is overhead. This alone would fool our brains into thinking the moon on the horizon is further away than normal and thus must have expanded.

❓ Why are all the planets spherical?

An object's shape is the result of the various forces at work upon it. For the planets, their huge mass makes gravity the most important force. By pulling together material equally strongly at equal distances from the centre of mass, gravity creates a spherical end-product – or as close to spherical as it can manage in the presence of other forces, including the inter-atomic forces within the material making up the planet, which resist the inward force of gravity and the effects of any rotation and heat. The end result is typically less than perfectly round. For example, the Earth's daily rotation produces an equatorial "bulge" that makes it around 27 miles wider around the middle than across the poles. In addition, there are bumps and dips left by the inability of gravity to overcome the inter-atomic forces within the rocky body of the Earth. The most obvious examples of these are mountain ranges and ocean trenches but larger-scale undulations have been revealed by studies of satellites, whose orbits are affected by the resulting changes in gravitational attraction. In December 1958, American scientists announced the discovery of a 50 ft "dimple" at the South Pole, implying that the Earth is actually a very fat, squat pear. Since then, probes have visited all our neighbouring planets except Pluto and carried out detailed studies of their shapes. These studies have confirmed that Saturn is the least spherical planet, its rapid rotation rate making it over 7,000 miles wider at its equator than its poles.

❓ Why are planets, stars and galaxies all rotating?

The principal cause is a combination of gravity and the conservation of angular momentum. As concentrations of cosmic dust and gas collapse under the action of gravity, even a slight swirl in the primordial material (due, perhaps, to turbulence) will – like an ice-skater pulling in her arms – be amplified to produce significant rotation in the final object. Gravitational close encounters and actual collisions can also create spin-inducing torques, especially in smaller, planet-sized objects. The Earth is thought to have acquired its spin after being struck by a Mars-sized object just after its formation.

However, while these effects account for the basic phenomena, explaining the fine detail is far more difficult. For example, if the Sun and the planets both condensed from the same mass of swirling matter, the "ice-skater effect" should have left the sun spinning much faster than it does; its leisurely 25-day rotation points to some subtle braking effect. More mysteriously still, a graph plotting the masses of planets, stars and galaxies against their angular momentum gives a neat parabola, for reasons never plausibly explained.

? Why can you sometimes see the "dark" half of a crescent moon?

This "earthshine" is caused by the moon being lit up by sunlight bounced off the Earth. The effect was first explained by Leonardo da Vinci almost 500 years ago and is most obvious near a New Moon, when the Earth is fully lit up by the Sun and thus reflects most sunlight on to its celestial companion. Earthshine is also stronger when there are more light-reflecting clouds in our atmosphere – which has led some scientists to use it to monitor changes in the Earth's climate.

? Why does the Moon always present the same face towards the Earth?

Despite appearances, the Moon spins on its axis once every 27.3 days. The reason we don't notice is because this matches the rate at which the Moon orbits the Earth, thus ensuring that the same face is turned always towards us (although some subtle effects actually allow us to peek around each side, and see fifty-nine per cent of the surface). This "synchronous rotation" is caused chiefly by the Earth's gravity distorting the shape of the spinning Moon as it travels around us, which applies a braking force such that the Moon is compelled to match its spin-rate to its orbital period. Virtually every moon in the solar system exhibits synchronous rotation but only one planet shows any signs of going the same way: Mercury, whose relative closeness to the Sun has slowed its spin-rate to one rotation every fifty-eight days. The Earth is so far away from the Sun that the braking force is extremely feeble and so there is no danger (or no hope, depending on your view) of its spin-rate eventually reaching the point where one side of our planet constantly faces the Sun.

? How likely is Earth-like life to exist elsewhere in the universe?

To judge by *Star Trek* or *ET*, the universe is teeming with lifeforms that look just like humans, apart from having pointy ears or a luminous index finger. If our universe is truly infinite in extent (and as the Roman philosopher Lucretius pointed out, what else can it be?), then it is absolutely certain that beings identical to us in every particular exist out there somewhere – though, by the same token, we are unlikely ever to encounter them.

Even a cursory knowledge of the evolution of life on this planet makes clear there is a vanishingly small probability of the same course of events being repeated elsewhere, at least in our neck of the cosmic woods. It would require the formation of a planet like ours, at just the right distance from a sun-like star and with a suitable mix of atmospheric gases. Then the biochemical processes that led to the emergence of DNA-based life would have to be repeated. As we still don't know what those processes were, this is hard to assess but it has been estimated that the simplest living cell requires at least fifty genes and that the odds of exactly the same sequence of DNA needed for these genes emerging again is one in 10 followed by 5,400 zeroes. Odds do not come much longer than this.

? Whatever happened to the idea of artificial gravity in spacecraft?

For those of us raised on *Dan Dare* and *2001: A Space Odyssey*, it was a racing certainty that the space stations orbiting above us would perform stately pirouettes to give their inhabitants a sense of gravity by the "spin-drier" centrifuge effect. Yet, here we are, well past 2001 and the astronauts aboard the International Space Station are still gormlessly floating around like helium balloons. Although very simple from an engineering point of view, the spin-drier approach plays havoc with human physiology. Experiments conducted during the 1960s and 1970s found that people start to feel queasy when exposed to spin rates greater than about two revolutions per minute. Such a slow spin rate means that a space station would have to measure around 1000 ft across in order to generate enough centrifugal action to mimic Earth-like gravity – far bigger than anything presently feasible. If humans are to explore space extensively, the problem of generating artificial gravity will have to be solved, as studies of astronauts on long space missions have shown that prolonged exposure to weightlessness weakens muscles and bones. One idea is to give astronauts occasional exposure to gravity in a man-sized onboard spin drier but the small size of the unit means that the rotation rate has to be around twenty revolutions per minute and experiments have shown nausea is a real problem. My favourite suggestion for generating artificial gravity is to propel the whole spacecraft forward at a rate of acceleration matching that generated by the Earth's gravity. Unfortunately, no one has yet devised a propulsion system capable of such prodigious performance.

❓ How is weightlessness created in an aircraft?

The trick lies in putting the aircraft in a dive that follows the same path as an object would if thrown off a cliff with the same horizontal speed as the aircraft. The shape of this trajectory is a partial parabola and in following it the plane and its contents behave as if they were in free-fall, everything falling towards the Earth with the same acceleration. Thus, as the crew and passengers of the aircraft are falling just as fast as the floor of the aircraft, they find themselves apparently floating in mid-air inside the plane. The sensation does not suit everyone, as NASA discovered when it used these parabolic flights to give would-be astronauts around thirty seconds experience of weightlessness: around a third of the passengers became violently sick. The aircraft used to make the flights came to be known (for all too obvious reasons) as the *Vomit Comet*.

❓ Why don't people south of the Equator feel they are upside down?

Our sense of up and down comes from the otolithic organs in the inner ear, each of which has a small sensory area, about 2 mm across, containing several thousand hair cells. The tiny hairs emerging from these cells detect any shift in heaps of chalky granules or *otoliths* (literally, "ear stones"). As the movement of these granules is affected by gravity, it is these that give us our sense of being upright or inverted relative to gravity.

❓ How and when did the Earth acquire the Moon?

The Apollo programme is often criticized for being more about beating the Russkies than contributing to human knowledge but the 800 lbs of moon rock they hauled back helped to cast crucial light on the origin of our nearest celestial neighbour. Before Neil Armstrong stepped on to the moon, there were several competing theories, ranging from the Earth simply capturing the Moon as it wandered by, to the Moon being spun out like a huge droplet when the Earth was still molten.

Chemical analyses of the lunar samples sounded the death knell for most of the theories, as they revealed striking similarities with the composition of the Earth's mantle. By the mid-1970s, a new theory was starting to emerge, in which the Moon was the product of a titanic impact between the Earth and some other planet-size body shortly after the formation of the Solar System. The impact was violent enough to gouge out a sizeable fraction of the Earth's internal material and the resulting debris then coalesced under its own gravity into the neat and tidy ball we call the Moon.

Working out the details, such as the likely size of the object that hit the Earth, had to wait until the advent of supercomputers powerful enough to simulate the complex effects of the impact and its aftermath. In the last few years, such simulations have led to the view that the Earth was struck by an object about the size of Mars and the resulting debris – around thirty per cent of it from the Earth – formed the Moon some time during the first 100 million years of the Earth's history. Studies of meteorites indicate that the planets were formed about 4,540 million years ago. Analyses of moon-rock, meanwhile, intimate that the Moon was formed between 4,500 and 4,520 million years ago. Taken together, these suggest the Earth has been orbited by the Moon for all but around 20–40 million years of its existence.

? Why are there two high tides each day?

There are also two low tides each day (to be exact, every 24 hours and 50 minutes, as a result of the Moon having moved a bit in its orbit). It seems reasonable to expect there would be just one of each in a day, as our part of the Earth comes round to face the Moon and thus experiences the full force of its gravity. This overlooks the fact that gravity does not just grab at whatever's closest: it affects everything, including the oceans round the other side of the world, albeit to a rapidly diminishing degree. It also ignores a subtle effect due to the proximity of the Moon, which causes the Earth and Moon to spin round each other, triggering a centrifugal force on the far side of the Earth. It's the combination of these two effects that leads to the formation of two ocean bulges. First, there's the "common sense" bulge on the face of the Earth closest to the Moon; then there's the second bulge on the other side, formed by the centrifugal effect and also by the diminishing influence of the Moon's gravity, which leads to the oceans there being "left behind" by the rest of the Earth. As the Earth spins, we run into both bulges, which we experience as high tides. I should add that, like anything to do with the Earth–Moon system, the details needed for accurate tidal prediction are much more complex (Sir Isaac Newton complained that such problems were the only ones that ever made his head ache). For a start, there is the role of the Sun, which boosts the Moon's pull during full or new moons to produce spring tides. Nor are the seas alone in being affected: twice a day, the Earth's surface rises and falls by around a yard at the Equator.

❓ How much is the earth affected by other planets?

While the planets may not affect us in ways astrologers would have us believe, events on Earth are undoubtedly influenced by the gravitational pull of our celestial neighbours and the results are rather more important than liaisons with tall dark strangers. The push and pull of other planets, primarily Jupiter and Venus, alter the size, shape and orientation of the Earth's orbit, while the axis of the Earth's orbit swings around under the joint effect of the Sun and Moon. Three of these effects are especially important. First, the eccentricity of the Earth's orbit – that is, its lack of perfect circularity – varies from almost zero to 0.12 per cent over a period of 100,000 years. Second, the tilt of the Earth's axis relative to its orbit also varies, from today's value of around 23.5° to as little as 21.6° and as much as 24.5°, over a period of around 41,000 years. Finally, the timing of the Earth's closest approach to the Sun is affected by the other planets, the Sun and the Moon. Today, the Earth comes closest to the Sun in early January, when the northern hemisphere is tilted away from the Sun. Around 11,500 years from now, however, the Earth's axis will have been pulled around, so that the northern hemisphere is tilting towards the Sun when the planet make its closest approach.

Apart from putting Christmas in the middle of the northern summer, these periodic changes, known as Milankovitch cycles, are believed to have profound effects on the Earth's climate, by altering the amount of solar heating that reaches the planet. The direct effects are very small but the sensitivity of the Earth's climate leads to dramatic consequences. Planet-induced Milankovitch cycles are now thought to play a role in the ebb and flow of the Ice Ages, and have recently been linked to the sudden change in the climate that led to the creation of the Sahara, around 5,500 years ago.

❓ Do meteorites make a noise as they fall to Earth?

Entering the Earth's atmosphere at speeds of 40,000 mph or more, most of these chunks of cosmic debris burn up too high to generate anything we can hear at ground level. Those that penetrate deep into the atmosphere create a shock wave of compressed air in front of them, which may reach the ground as an audible sonic boom. In many cases, the aerodynamic stresses become so great the meteors detonate in mid-air, with dramatic effect. In 1999, scientists at the seismology division of the Royal Netherlands Meteorological Institute detected the low-frequency infrasound released by a meteor that detonated at an altitude of around 15,000 m with the violence of a small atomic bomb. Occasionally these air bursts can be heard directly, as the inhabitants of a remote part of New Zealand's North Island discovered on 8 July 1999 (there are videos online). Very occasionally, meteors make it all the way down to Earth and mark their conversion to "meteorites" by hitting the ground with a mighty thump. Sadly, there is rarely anyone around to hear it and the arrival of another precious source of free inter-planetary material goes undiscovered.

❓ What is the furthest object visible to the naked eye?

This is one of those trick questions, usually posed as "How far can you see with the naked eye?". You might think it's around 100 miles, say from the top of Everest or some such high spot. The actual answer depends on how good your eyesight is but is a lot further than you might think. Anyone with moderate eyesight can make out the galaxy in Andromeda, which is 2,200,000 light-years, or about 13 billion billion miles away. Those with really good eyesight may even be able to top that by glimpsing the M33 galaxy in Triangulum, which is just visible in really clear skies and is around half as far away again, at 3,000,000 light-years.

❓ When an asteroid approaches the Earth, why doesn't gravity pull it in and cause a collision?

The Earth had a very close shave in 2002, when a 350 ft wide asteroid, 2002MN, came within 75,000 miles. If it had hit, the impact would have been as devastating as the explosion of a dozen or so H-bombs. In everyday terms, 75,000 miles seems like a huge distance but the Earth's gravity has some influence much further into space (witness our Moon, which is three times further away). The reason 2002MN managed to slip through the Earth's clutches is that it was travelling so fast rela-tive to the Earth – around 23,000 mph – that it had enough energy to climb back out of the Earth's "gravitational well" and off into deep space. A quick calculation shows that 2002MN would have had to come within 2,000 miles of the surface of the

Earth to be sucked in for a collision. Most asteroids have even higher speeds and would have to get within 1,000 miles of the Earth to end up hitting us. Still, it's only a matter of time before one does.

? How are the precise times of sunrise and sunset determined?

Given that astronomy is the oldest science and that sunrise and sunset are very basic celestial phenomena, you might have thought this would have been sorted out centuries ago. Not a bit of it. We live on a rocky ball surrounded with air, which makes it hard to pin down the time of sunrise and sunset with reasonable accuracy. For example, which part of the sun should be peeking over the horizon and which horizon: a fictitious, idealized flat plane or one taking into account local conditions like valleys and mountains? Furthermore, we can only see the sun's disc after its light has travelled through the Earth's atmosphere. The resulting refraction means the sun is not quite where it appears to be, an effect which varies with latitude and meteorological conditions.

The best astronomers have come up with is a definition of sunrise or sunset as the moment when the centre of the sun's disc makes an angle of 90.8333° with the zenith, that is, the point directly over the observer's head. This dispenses with the tricky problem of defining a standard horizon but why is the angle not exactly 90°? First, over the course of a year, the sun's disc appears to cover an average of 0.53333° of the sky, so the distance between the centre of the sun and its edge is half this, or 0.26667°. Second, the earth's atmosphere bends the light of the sun by an average of around 0.56666° and this must be added to the radius of the sun to ensure appearances match with reality, which gives the 0.8333 part of the final number.

The definition is, then, a bit of a fudge, based on annual averages and ignoring local circumstances. So, sunrise and sunset times can't be precisely predicted for any given locality (especially at relatively high latitudes, where refraction effects are strong) and the values quoted in newspapers can be out by a minute or more.

? Is there a single case of a spacecraft being wrecked by "space debris"?

Claims that the International Space Station could be destroyed by the impact of an orbiting metal bolt do have some basis. The energy involved in such a collision is truly fearsome. In low Earth orbit the average collision speed is around 22,000 mph, which will give even a small bolt the destructive punch of a car doing around 150 mph. Nor is there any shortage of objects capable of delivering a nasty impact. Astronomers estimate that there are currently around 100,000 bits of metal between half an inch and four inches across floating around in Earth orbit.

Happily, the threat from space debris has failed to make much of a, well, impact – so far. The Soviet Union suspected that the mysterious disappearance in 1981 of its Kosmos 1275 satellite was the result of an impact with space debris but the first confirmed case was in July 1996, when a French spy satellite was hit by a stray bit of metal travelling at 31,000 mph. Named Cerise, the 100 lb satellite had been launched the previous year to eavesdrop on the electronic communications of foreign governments. Its assailant was a cereal-packet-sized chunk of an Ariane booster rocket launched in 1986. Amazingly, despite its stabilization system being vaporized by the impact, Cerise's controllers were able to patch it up and continue the mission. To date, the Americans don't think they've lost a single spacecraft to space debris.

While these one-off events may not pose much of a threat, there is more concern about the "cascade effect", in which an impact destroys one spacecraft, creating lots of debris that then destroys other spacecraft and so on, in a devastating chain reaction. There have been proposals to avert this potential catastrophe by sending up lots of little satellites to tidy up our cosmic neighbourhood. So far, such plans have generated all the enthusiasm usually associated with tidying up.

❓ Where do the Space Station astronauts get their air?

Astronauts first moved into the International Space Station over four years ago and since then they have been supplied with oxygen created by a process discovered by the chemist William Nicholson over 200 years ago: electrolysis.

Water is transported up from the Earth to the ISS, where a Russian-built unit, the Elektron, uses electricity from the station's solar panels to split the water molecules into oxygen, which is fed into the station and hydrogen, which is expelled into space. To provide a safe, Earth-like, atmosphere and maintain a decent level of air pressure, the oxygen is combined with relatively inert nitrogen from on-board tanks.

While the basic technology may be pre-Victorian, the ingenuity comes in squeezing the very last gasp of oxygen from the water: some ninety-eight per cent is recycled – right down to the tiny amounts of sweat and water vapour in the astronauts' breath.

In addition to Elektron, the ISS carries emergency sources of oxygen, including several months' worth in tanks and over 100 lithium perchlorate candles, each of which, when lit, releases enough oxygen to keep an astronaut alive for a day.

❓ Is the moon moving away from us?

By bouncing laser beams off special mirrors left on the moon by Apollo astronauts, astronomers have been able to show that our nearest celestial neighbour is moving away at about 1.5 inches per year. This is believed to be a quid pro quo for the tidal drag that the Moon exerts on our planet, slowing down the rotation of the Earth and forcing the Moon further away from us to make up for the lost angular momentum.

There will eventually come a time when the Moon has moved so far away that its disk will appear too small to cover that of the Sun, robbing us of a spectacular celestial phenomenon unparalleled in the solar system: total eclipses. Assuming the actual physical sizes of the Moon, the Sun and our orbit about it don't change, a back-of-the-envelope estimate suggests that, at the current rate of recession, the Moon will no longer completely eclipse the Sun around 420 million years hence. So there is no immediate need to panic.

Things are rarely simple in celestial mechanics, so I raised the issue with Dr Duncan Steel, astronomer and author of the outstanding *Eclipse* (Headline Press, 1999). He points out that the guesstimated timescale is vast – so vast, that it's possible some of the assumptions may be wrong. For example, the Sun's diameter may well change significantly over 400 million years, as a result of burning through a sizeable fraction of its nuclear fuel. The size of the Earth's orbit may also have changed, in a way that could cancel out the change in the sun's diameter. The details are not easy to estimate and Dr Steel adds that, in any case, solar eclipses will cease, at least temporarily, long before 400 million years have elapsed, as the result of periodic changes in the orbits of both the Moon and the Earth. Whatever: for most of those living in the UK today, total solar eclipses are already history: none will be visible here until 2090.

❓ Is Pluto really a planet?

Ever since its discovery in 1930, Pluto's status as the ninth and outermost planet of the solar system has been questionable. It has an orbit far more eccentric and tilted than any other planet and is by far the smallest, with a mass barely one-sixth that of our Moon. To my mind, the clinching evidence against Pluto's status has been the discovery, over the last decade, of a host of "Kuiper Belt Objects". These chunks of ice and rock orbit at the edge of the solar system and look just like mini Plutos. In 2002, astronomers found one KBO with a diameter half that of Pluto. It may well be just a matter of time before a KBO bigger than

Pluto turns up – in which case, we should do the decent thing and demote this celebrity snowball.

? Why are the inner planets rocky while the outer ones are balls of gas?

How the solar system came to take the form it does is one of those questions that refutes the notion that the more we know, the clearer things become. As long ago as 1755, the philosopher Immanuel Kant put forward the idea that the sun and planets condensed out of a huge cloud of dust and gas. Various bells and whistles have since been added to this "nebula hypothesis", which offers a nice, hand-waving, solution to the question of why rocky little planets like the Earth and Mars are closer to the sun than galumphing "gas giants" like Jupiter and Saturn. Quite simply, the presence of the hot sun stripped the inner regions of the primordial cloud of volatile gases, leaving behind just the heavier stuff that collapsed under its own gravity to form the rocky planets, one of which we now inhabit.

As well as accounting for our own solar system, this theory makes a prediction: that Jupiter-like planets should never be found very close to their parent stars in other solar systems either. Yet as Professor Scott Tremaine, a planetary scientist at Princeton University, recently ruefully observed: "Most every prediction by theorists about planetary formation has been wrong" and this one is no exception. No sooner did astronomers begin finding planetary systems beyond the solar system than they found huge gas giants orbiting very close to the central stars.

They've tried to come up with some alternative explanations, such as the notion that gas giants may indeed be formed in the colder parts of the nebula but then spiral inwards, but each has its problems (for example, explaining why the gas giants don't just carry on spiralling in until they burn up). It may just be that there isn't a simple explanation. This happens sometimes.

? Why can't nuclear waste be sent off to the sun aboard rockets?

The UK alone is sitting on over 3,000 tons of high-level nuclear waste and getting rid of it safely is a pressing problem. On the face of it, sending it all off to the Big Incinerator in the Sky seems an excellent idea; with a payload capacity of 25 tons, the Space Shuttle could rid the UK of its high-level waste in around 120 flights, roughly the tally of flights to date. The reason it's not being done is because an accident would shower high-level radioactive waste across vast areas – and as the 2003 Columbia disaster showed, this isn't just a theoretical possibility.

? How do scientists know that the famous "Martian bug" meteorite came from Mars?

I can still recall where I was when, in July 1996, I heard of the amazing discovery by NASA scientists of "bacteria" in a meteorite from Mars (in a newsagent's in Aberdovey, as it happens). I can also recall my immediate reaction being disbelief: how did they know the microscopic, blob-like things on the meteorite were bacteria and, moreover, bacteria from Mars? Sadly, my scepticism proved at least partly well-placed. The current consensus is that the evidence that the blobs are fossilized bacteria is far from compelling, subsequent studies having shown that their characteristics could have been created by lifeless chemistry. There is, however, little doubt that the meteorite itself came from the Red Planet. Indeed, it's remarkable just how much scientists can say about the history of meteorite ALH 84001.

Its birthplace on the surface of Mars was established in 1995 by analysis of the tiny amounts of gas trapped in its tiny spaces: hydrogen, nitrogen, argon and xenon: its unusual mix of isotopes of these elements is unlike any known on Earth but matches that found in the atmosphere of Mars by NASA's Viking probes, which landed on the planet in 1976.

Using standard dating methods based on radioactive isotopes, the artefacts are estimated to be at least 1,400 million years old. It's even possible to estimate when the meteorite left the surface of Mars and how long it took to come to Earth. On its way from Mars, ALH 84001 would have been exposed to cosmic rays, creating various isotopes within it at a rate tied to the length of its interplanetary voyage. The data are consistent with a journey time of around 16 million years, with other isotope measurements suggesting it landed in Antarctica around 13,000 years ago. As for how ALH 84001 came to leave Mars in the first place, the best guess is that it was ejected in the aftermath of the impact of an asteroid.

? Why do the planets in the solar system all spin round and orbit the sun in an anti-clockwise direction?

Actually, Venus, Uranus and Pluto rotate clockwise, possibly as the result of being struck and flipped over by impacts early in the history of the solar system. Still, the rest rotate anti-clockwise and certainly they all orbit the sun in an anti-clockwise direction. Explaining why requires answers to two questions: why do the planets all go round the sun in the same way and why anti-clockwise? The answer to the first lies in the process by which the solar system was formed. The prevailing view (though one not without its problems) is that the Sun and planets formed from a cloud of dust and gas that collapsed under its own gravity; in the process, it formed a disc that rotated ever faster and contained regions of somewhat higher density, which in turn collapsed to form "planetesimals": small aggregations of gas and dust that clumped together to form the planets.

With their common origin, it's to be expected that they all share the same orbital direction – but why anti-clockwise? The answer lies in the definition of "clockwise"; the direction in which the fingers of a clock rotate. This was itself defined centuries ago by the clockwise movement of the sun across the sky and thus of the shadow round sundials. Of course, the sun doesn't really move clockwise across the sky: it is the very fact that the Earth rotates the other way that creates this illusion and in the process dictates the direction of planetary motion as "anti-clockwise".

? Why can't the space shuttle re-enter the Earth's atmosphere gently, using its rocket engines to slo it down the whole way?

It's a matter of safety and cost. It takes a colossal amount of energy to put the Shuttle into orbit, as the spectacular blast-offs demonstrate. That energy has to be destroyed again by the time the Shuttle returns to base, which certainly *could* be done using the retro-rockets' power. The trouble is, this would take roughly as much fuel as was needed to get into orbit in the first place, which means taking up yet more fuel at colossal expense. Then there's the unnerving possibility of the retro-rockets failing on the way down. Using aerodynamic drag to shed energy is both much cheaper and much safer – as attested by the fact that the 2003 Columbia failure was the first heat shield failure in forty years of space flight.

? How is the gravity of planets used to boost the speed of space probes?

It would seem obvious that, while a probe will gather speed as it approaches a huge planet like Jupiter, it will lose it all as it departs, producing no net advantage. That is what some of the world's leading experts on space flight told the twenty-some-thing student Gary Flandro, when he came up with the "gravitational slingshot" idea in the mid-1960s. As he succinctly put it in *Planets Beyond: Discovering the Outer Solar System* by Mark Littmann (Wiley, 1990), the experts had forgotten that the planets move around the sun. The trick to the slingshot lies in approaching Jupiter from behind, so that its gravity pulls the probe in and tries to compel it to follow it around the sun. The probe then speeds up, as a result of the pull of the planet's gravity and being caught up in its motion around the sun. As it swoops behind and away from Jupiter, the probe will be slowed down by the planet's gravity but still benefits from the energy it extracted from Jupiter's motion around the sun and zooms on to its next destination. It is just as well Dr Flandro was able to convince the experts quickly, as an unusual alignment of the planets in the late 1970s meant that his ingenious idea would allow probes to visit all the outer planets in one "grand tour". The result was the spectacularly successful Voyager missions, which exploited a window of opportunity that will not open again for well over a century.

? Is there really a Planet X beyond the known planets?

The possibility of there being more than nine planets in our solar system was raised shortly after the discovery, in 1930, of Pluto, currently the furthest known planet. The search for Pluto had been prompted by discrepancies in the calculated orbit of Uranus, which suggested it was being pulled off course by another, unknown mass. When Pluto was discovered, however, it proved to be far too small and distant to account for the discrepancies – leading some astronomers to speculate about the existence of a larger "Planet X", even further away. The idea was dealt a major blow in the late 1980s, when the Voyager 2 probe sent back new values for the masses of the outer planets. When the orbits of the planets were recalculated using these data, the discrepancies that had prompted belief in Planet X vanished.

? Can any telescope see a star's disc?

Seen through even the most powerful telescope, every star beyond the sun looks just like it does through a toy spy-glass: a single point of light. In 1868, the physicist Armand Fizeau pointed out that the width of a star could still be measured by exploiting the wave properties of light. The trick lies in using a pair of mirrors to combine the rays of light coming from the edges of the star. By moving the mirrors so that the light rays interfere, it's possible to work out how far apart the rays originally were and thus how big the star is. In 1920, the astronomer Albert Michelson made the front page of the *New York Times* after using the method to reveal that the red giant star Betelgeuse is 259 million miles across or 300 times the Sun's diameter. For decades, this seemed to be the limit of the 'interference' technique but during the 1980s Dr John Baldwin and his colleagues at Cambridge University began studies that turned it into a means of seeing stars' discs. The details are complex but, in essence, extra mirrors are used to fill in the rest of the star's image, thus revealing the whole disc. In September 1995, the team obtained images of the disc of Capella, a bright star in the northern constellation of Auriga and confirmed previous evidence that it is really two stars orbiting one another. The team has since obtained spectacular images of several other stars, including Betelgeuse.

❓ Why is Saturn the only planet surrounded by rings?

While Saturn's rings are certainly spectacular, they are not unique. A dark ring system was found around Uranus when, in March 1977, the planet passed in front of a faint star. The light from the star faded several times as it passed to either side of the planet; later observations revealed the presence of eleven separate rings. A set of three rings was found around Jupiter in 1979 by the Voyager 1 spacecraft and four around Neptune by its sister craft when it visited in 1989.

The origin of Saturn's rings has been a puzzle ever since they were first glimpsed by Galileo in July 1610. Through his relatively small telescope, the rings looked like two blobs stuck on either side of Saturn; it took the more sophisticated telescope of the astronomer Christiaan Huyghens to reveal the rings in 1655. In 1859, the brilliant Scottish theorist James Clerk Maxwell proved mathematically that the rings could not be solid, as they would be torn apart by the internal stresses caused by Saturn's gravitational field. His prediction was not conclusively confirmed until 1979, when NASA's Pioneer 11 probes revealed the rings to be made up of ice-like particles no more than a few metres across. How they got there is still a mystery: the best guess is that a wandering comet passed too close to Saturn and was shredded by its gravitational field.

? Why do only stars twinkle, and not the moon and planets?

Twinkling – scintillation, to use the scientific term – is the result of turbulence in the Earth's atmosphere, which causes random jitters in the rays of light from celestial objects. The size of these jitters varies with local conditions, such as heat rising into the night sky from towns but even at the best observatories on top of mountains is still far larger than the point-like images of stars. This makes them seem to dance around or, as we call it, twinkle. The Moon and planets are far closer and thus don't appear as mere points of light, so the jitters are relatively small and don't have so pronounced an effect. Seen through a telescope, the effects of the turbulence can be seen as a quivering around the edge of the disc of the moon and planets. One of the biggest advances in astronomy in recent years has been the development of "adaptive optics" which cancels out the effect of turbulence. It's done by constantly flexing the mirror of the telescope under computer control to produce images as good as if they were taken using a telescope in space (as a bonus, it's also about a billion pounds cheaper).

Chapter 9

Cosmic conundrums

? **What causes the pull of gravity?**

Of all the science questions I have been asked over the years, this is the most common, suggesting many people are as baffled by gravity as was Sir Isaac Newton. You might think that the formulator of the Law of Universal Gravitation would have had a grip on the subject but not at all. Newton's public stance, set out in 1713 in his most famous book, the *Principia*, was that "I have not been able to discover the cause of those properties of gravity from phemenon, and I frame no hypotheses ... to us it is enough that gravity really does exist". So there. Privately, Newton seems to have regarded gravity as a manifestation God's all-pervading spirit, which he knew would cut little ice with his scientific rivals. The key mystery is how a mass like the Earth can affect distant objects through the vacuum of space. Einstein supplied the answer in 1915, in his General Theory o Relativity. This viewed gravity as distortions caused by mass in the very fabric of space and time. It is a very deep idea, as all our perceptions of motion and force are intimately tied to our view of space and time as being unchangeable. Einstein says we must give this up and think of gravity as curving space and time in such a way that it creates the impression of objects being attracted to each other. Roughly speaking, Einstein is saying that we believe gravity is a force because our measurements of the fall of an apple are consistent with the characteristics we expect of one. What's actually happening, however, is that our measurements of space and time have been distorted by gravity to create those characteristics.

❓ What is the speed of gravity?

Sir Isaac Newton believed that gravity travelled across space infinitely quickly, presumably on the basis that it was hard to imagine at what finite speed a God-given universal force could travel. It wasn't until Einstein published his General Theory of Relativity (GR) that it became clear that gravity does have a speed. According to GR, gravity is the result of masses distorting space and time around them in such a way that it seems to act like an attractive force. Einstein was able to extract a wave equation from his theory, which showed that violent motion by matter sends out ripple-like distortions of space-time. This wave equation, published in 1916, predicts that these ripples travel at the speed of light – 186,000 miles a second – which can thus be taken as the speed of gravity. Thus, if the Sun were to vanish right now, it would take around eight minutes for the waves of gravity generated by its disappearance to reach us.

That's the theory: to this day, however, no one has confirmed it to the satisfaction of most scientists. In January 2003 Professor Sergei Kopeikin, of the University of Missouri-Columbia, created a stir by claiming to have measured the speed of gravity using measurements on light rays from a distant galaxy passing through the gravitational field of Jupiter. The result appeared to show that the speed of gravity matched that of light to within twenty per cent. Unfortunately, few physicists seem to have much faith in the basic theory behind the experiment, so its success is not widely regarded as confirmation that Einstein was right yet again.

❓ How fast are we moving through the universe?

There was a time when such questions were routinely batted away by invoking the hallowed name of Einstein and declaring "It's all relative". Fortunately, things have now become more interesting, following a truly cosmic discovery. The fact that we are moving even when standing still is apparent from the slow movement of the stars during the course of an evening and during successive nights. Yet, as well as standing on a planet moving at 67,000 mph relative to the Sun, our solar system is travelling around the galaxy at around 500,000 mph, which itself is moving towards another set of galaxies at around the same speed. There is an ultimate backdrop against which we can measure our speed; the heat left over from the Big Bang. Discovered in 1964, this permeates the whole of space and any movement relative to it can be detected by very precise temperature measurements. In 1996, a team led by Dr Dale Fixsen at the Goddard Space Flight Centre, Maryland, used this method to show that the Earth is moving at a speed of around 830,000 mph relative to the universe as a whole and in the direction of a point located on the border of the constellations Leo and Crater.

❓ What would objects look like if we could travel near the speed of light?

It is often said that the teenaged Einstein was inspired to develop his theory of relativity by asking himself what it would be like to ride on a beam of light. Rather less well-known is that not even he appreciated how weird the experience would be. His theory predicts that objects approaching the speed of light shrink in the direction of motion and you might expect this shrinking to produce visible effects. However, in 1959, the celebrated mathematician Roger Penrose showed this wasn't always the case. A sphere whizzing by at near light speed would still appear circular but a rocket doing the same would appear squashed and slightly twisted around, allowing bits of it not normally visible to come into view (a phenomenon known as Penrose-Terrell Rotation). The scene would be no less bizarre for those inside the rocket. They would see objects not only shrink but also bend out of shape, while their brightness and colour change, appearing blindingly blue-white as they approach and dull red-to-black as they recede into the distance.

? How did "nothing" manage to turn itself into "something"?

Questions do not come much deeper than this and until the 1970s only theologians and philosophers felt able to take it on (plus physics students still in the bar at closing time). The first glimmerings of what has become the standard scientific response emerged in 1973, in a now-famous paper in the journal *Nature*, "Is the Universe a Vacuum Fluctuation?". Its author, the physicist Edward Tryon, pointed out some curious features of our universe. For example, the total amount of positive energy locked up in cosmic matter matches (roughly speaking) its total gravitational energy, which is negative. In other words, the total amount of energy in the universe appears to be zero. Together with various other more technical properties, this led Tryon to suggest that the universe was created from nothing or, more precisely, what physicists call the vacuum state. The beauty of Tryon's suggestion is that the properties of this vacuum state can be calculated, so the question of how "nothing" became "something" was opened up.

Since then, physicists have become convinced that the origin of the universe is inextricably linked to the properties of the vacuum state, from which it burst forth in a Big Bang around fourteen billion years ago. Such spontaneous creation might seem like the ultimate conjuring trick but is entirely plausible. Laboratory measurements have long established the fact that sub-atomic particles are constantly popping in and out of the vacuum state all around us. The spontaneous appearance of something the size of the universe is less frequent of course but, crucially, is not impossible. If Tryon is to be believed, that is precisely what has happened.

Not surprisingly, not everyone is convinced – not least because Tryon was a bit cavalier in his definition of the universe. Strictly, it is not merely the things in it but also space and time as well and quite how these emerged out of Nothing is still very controversial.

My personal view is that fundamental physics just isn't fundamental enough to explain how Nothing became Something and that we should look to pure mathematics for answers. Many years ago, the mathematician John von Neumann showed it was possible to conjure up Something from Nothing via set theory; that is, the properties of collections of things. In particular, von Neumann gave an ingenious proof that all the numbers we count with, from zero onwards, can be created from arrangements of the "Empty Set", which contains nothing at all. I used to take this as a hint that mathematics held the key to how Something arises from Nothing. Now I suspect it is telling us something rather different: that in reality Nothing is all there is and it is our insistence on packaging this up into finite sets that creates the impression that it really is Something.

❓ How big is the universe, and how fast is it expanding?

This question begs a whole lot of others, not least because the universe is supposed to be all there is, so surely its size is infinite. None the less, it does make sense to talk about the "size" of the universe, because of the very specific way in which it expands. The universe, on its very grandest scale, appears to be pretty uniform, even titanic clusters of galaxies seeming like tiny dots lost in the abyss of space. This being so, one can show that the further one looks into space, the faster distant galaxies in the universe will appear to recede from us. This is summed up in Hubble's Law (named after the astronomer who discovered it observationally in the 1920s), which states that on truly cosmic scales galaxies R light-years away from us will appear to be racing away from us with velocity v, given by the formula $v = HR$, where H is "Hubble's Constant". From this one might think the radius of the universe is the distance where objects appear to recede from us at the speed of light, in which case we will never be able to see them, as their rays can never reach us. In reality, things are more subtle. For a start, Hubble's Constant isn't a time constant. Even so, we can get a rough order-of-magnitude estimate of the "radius". First we need a value for the Hubble Constant: the current best estimate of this is around 50,000 mph per million light-years; that is, for every million light-years further into space we look, galaxies will appear to race away from us at an additional 50,000 mph. Plugging this into Hubble's Law, with v equal to the speed of light, the radius of the universe turns out to be around 13,600 million light-years; the true figure is about three times bigger.

? When we look into space, why do we not find ourselves looking at the Big Bang?

As light takes time to reach us, looking into space is equivalent to looking back in time, so we might expect that if we look beyond the most distant galaxies, we should eventually see the Big Bang. The reason we can't is partly because of events that took place in the very early universe and partly because we don't possess the right kind of vision. Following the Big Bang, around fourteen billion years ago, the universe was filled with a mixture of sub-atomic particles and radiation, which was very effective at scattering light, rather like smog. This cosmic pea-souper persisted for around 300,000 years and prevents us from directly viewing earlier events, including the Big Bang.

Once the universe cooled sufficiently to allow sub-atomic particles to form atoms, the fog cleared and radiation stood some chance of reaching us. The reason we can't see it directly is that on its way to us this radiation has been stretched by the expansion of the universe, increasing its wavelength towards longer, redder parts of the spectrum. Calculations show that it set out as infra-red radiation and has since been stretched by a factor of around 1,000 into microwave radiation, which our eyes cannot detect, but which can be "seen" by radio telescopes; indeed, the prediction of this "cosmic microwave background" and its subsequent discovery in the mid-1960s, is the best evidence for the reality of the Big Bang.

Having said it is impossible to detect radiation until 300,000 years after the Big Bang, I should add that astronomers are developing instruments that might be able to detect the Big Bang itself. These are "gravitational wave detectors", which may one day detect the ripples of space and time triggered during the birth of the universe.

? If the universe is expanding, why is the Andromeda Galaxy moving towards us?

The Andromeda Galaxy is around 2,200,000 light years away and has the distinctions of being both the most distant object easily visible to the naked eye and the only spiral galaxy heading towards us. According to current estimates, it will collide with our Milky Way in around 3,000 million years. So why isn't the supposed expansion of the universe able to spare us this fate by keeping Andromeda away? The answer is that the cosmic expansion does not affect everything on every scale (indeed, if it did, we would not be able to detect it, as our instruments would expand at the same rate). Roughly speaking, the cosmic expansion only starts to make its presence felt on scales at least 100 times larger than the distance between us and Andromeda. Below that, the force of gravity is dominant and astronomers think that our galaxy and Andromeda are spiralling in towards each other under their mutual gravitational attraction.

? Is there an ultimate maximum temperature?

As every schoolchild knows (well, used to know anyway) there is an ultimate minimum temperature, Absolute Zero – 0 kelvin or –273.15 °C. At this temperature, atoms are static apart from a tiny irreducible amount of jittering due to quantum effects. In 1966, the theoretical physicist (and Soviet dissident) Andrei Sakharov pondered the question of whether there could also be a maximum temperature. He argued that it would be linked to the maximum amount of radiant energy that could be crammed into the smallest volume of space. Quantum theory shows that there is such a minimum volume, set by the fact that, at some incredibly small scale, the very meaning of "space" no longer makes sense. Quantum effects that are normally negligible become dominant at this scale and make it impossible to say where one bit of space ends and another begins. This weird state of affairs emerges on scales of around 10^{-35} m – far smaller than even a sub-atomic particle. This in turn leads to a minimum conceivable volume of around 10^{-105} m^3. Sakharov came up with a similar argument to estimate the maximum amount of energy that can be crammed into this tiny volume and then worked out the temperature of the resulting radiation. It comes out at approximately 10^{31} °C. This is far beyond any temperature created by humans, the record for which is around 10^{17} °C, reached inside large particle accelerators. Indeed, such temperatures have only existed once before, during the birth of the cosmos around 14,000 million years ago.

? What was the original size of the universe at its birth?

All statements about the early universe stem from working backwards from its observed expansion. This implies that the universe was once much smaller; the question is, just how much? As far as physicists can tell, there is only one limit on the size of the early universe, set by the "Planck Scale" (after

Max Planck, the founder of quantum theory). This is very small: around 10^{-35} m, so small that protons would seem the size of entire galaxies. Calculations suggest that, at these inconceivably small scales, the processes capable of triggering the Big Bang could have taken place. Not that anyone knows for certain – such calculations push current physics to its breaking point and possibly beyond.

? If the universe is expanding, where is the force coming from?

In January 1998, astronomers unveiled evidence that the universe is not merely expanding but expanding at an ever-increasing rate. The claim, based on studies of the light from distant exploding stars, has since been backed by data from other sources, convincing most astronomers that it's genuine. This implies there is "anti-gravitational" force at work in the universe, propelling the cosmic expansion against gravity. Most theorists think it's due to sub-atomic effects that take place even in the vacuum of space. The big problem is that, so far, all attempts to calculate the strength of this cosmic force have proved laughably wrong: around a factor of 10^{110} higher than it can possibly be. Nobel Prizes are on offer to those who can do better.

? Where is the centre of the universe?

Like most questions about cosmology, it's hard to imagine how the universe can be expanding in all directions but not have a centre. One way to picture what's happening is to imagine the three dimensions of space being squashed on to the two dimensional surface of an uninflated balloon, with galaxies represented by coins stuck to the surface. Picture yourself sitting on one of these coins as the balloon is blown up. Looking around, every other coin appears to be moving away from you; it seems like you are at the centre of the expansion but clearly this is not so – whatever coin you sit on, you will see all the other coins racing away from you. No matter where you move to on the balloon's surface, you'll never find the "centre" of the expansion, because there isn't one: everywhere is moving away from everywhere else. It is just the same with the real universe except that everything takes place in extra dimensions, so it's well-nigh impossible to visualize.

? Where did the Big Bang take place?

Perhaps the single hardest concept to grasp in cosmology is the definition of the universe as the totality of everything: not just matter but also space and time. Human beings are adept at understanding everyday phenomena, which take place within the arena of space and time and all statements about the universe include that arena. Thus, in terms of what humans can perceive, the question of what the universe is expanding into is, quite literally, meaningless. An analogy may be helpful here. Imagine two-dimensional creatures crawling over the surface of an expanding balloon. While they can tell that the balloon is increasing in size (as they are getting ever further apart), their two-dimensional nature prevents them from ever perceiving what the balloon is expanding into. Humans, on the other hand, by virtue of our three-dimensional nature, can see that the balloon is expanding into a third dimension. Similarly, in the case of the universe, only a higher-dimensional being can make sense of the question of into what our universe is expanding. We mere humans must content ourselves with a mathematical understanding.

? How many atoms are there in the universe?

Physicists pride themselves on being able to make rough estimates of anything, even the number of atoms in the universe. Strictly speaking, it would be an estimate of the number in the visible universe; that is, roughly speaking, up to the point at which the rate of cosmic expansion equals the speed of light, so we can never know what lies beyond it. Estimating the number of atoms is then a matter of estimating the total mass of atoms within this cosmic sphere, and dividing by the mass of a typical atom. Easy, really: the best estimates for the various quantities required come from the results from NASA's WMAP probe, released in February 2003. These show that the radius of the visible universe is 13.7 billion light years and that the universe contains an average of just one atom per 5 cubic yards of empty space. So, by my calculations, that means there are around 2×10^{78} atoms in the visible universe.

? Does anyone know what atoms and molecules actually look like?

Astonishingly, the very existence of atoms was still being hotly debated as recently as the early twentieth century. Many leading scientists regarded atoms as a fiction that helped form mental pictures when trying to understand the properties of matter. As there was no possibility of actually seeing the things (not in those days, at any rate) those of a logical positivist turn of mind regarded atoms as non-existent. We know that the naturalist Robert Brown unwittingly saw indirect evidence for the reality of atoms as early as 1827, in the random jigglings of specks inside pollen grains. "Brownian Motion" is the result of otherwise invisible atoms hitting the far larger specks, making them dance around in a random fashion (as was first pointed out a century ago by a young Swiss patent clerk, Albert Einstein). The invention of the electron microscope in the 1930s allowed atoms to be photographed but the fuzzy blobs seen in such photographs are hard to square with the mental image most people have of the atom as a miniature solar system, with a collection of particles at the centre with electrons whizzing around it like tiny planets. Quantum theory tells us that atoms are far less concrete entities than this and that any mental image we might have cannot hope to capture the reality. Instead, scientists have been forced to settle for picturing atoms in various ways, depending on the circumstances. None of them is "right"; at best they only capture some facet of reality. But of the reality of atoms – whatever they look like – there can be no doubt at all.

? As atoms are mostly empty space, why does matter seem so solid?

This is one of those simple questions with a surprisingly deep answer. The apparent solidity of everything around us is a bit perplexing. Consider: atoms consist of a pretty insubstantial cloud of electrons whizzing around a nucleus barely 1/50,000 times the width of the overall atom. In other words, the apparently solid ground beneath us consists of 1 part matter to 100 million million parts empty space. The situation is even more curious, as the attractive force between the atoms that make up matter should have led to everything collapsing in a trice. The fact that we are here to ponder this question shows that nature has some way of preventing it happening. The answer to this difficulty did not emerge until the mid-1960s, when the theorist Freeman Dyson showed that, if atoms become too closely packed, their clouds of electrons start to interact with each other, creating a repulsive force. There comes a point where this outward "degeneracy pressure" exactly balances the attraction between the atoms and matter becomes stable. Estimates for the strengths of these two effects suggest that this balance point comes when solids are a few times denser than water – which is just what we see.

? What would happen if you fell into a black hole?

Black holes are notorious for having gravitational fields so intense that not even light can escape from them. So, you might think that approaching them would generate a sense of increasing heaviness or of being crushed. The reality could hardly be more different. Those unfortunate enough to be sucked in towards a huge black hole, such as the four million mile wide leviathan thought to lurk at the centre of our galaxy, would feel no force at all, as they would be in free fall. Even as they crossed over the "event horizon" – the "surface" of the black hole, marking the point at which not even light can back out again – they would feel nothing especially odd. Approaching the centre, however, the difference between the strength of gravity acting on their feet and heads would start to rise because their feet would be slightly closer to the source of gravity at the heart of the black hole. These "tidal forces" would rapidly increase, stretching everything lengthways while simultaneously crushing them sideways. In a matter of seconds, a human body would be turned into a strand of ultra-thin, ultra-long spaghetti and then shredded.

Oddly enough, things would be even worse with a small black hole such as the five mile wide objects thought to be left behind by collapsing stars. Being more compact, their tidal forces are far stronger and anything approaching them would be torn apart even before it crosses the event horizon.

? Does the universe contain "white holes" as well as black ones?

As its name suggests, a white hole is the exact opposite of a black hole. Where a black hole is a region of space into which matter disappears like water down a plug-hole, a white hole would allow matter to burst out of nowhere, like a water fountain. There's no question that white holes can exist, at least mathematically speaking. The starting point could be the collapse of a huge star once its nuclear fuel has run out. With nothing to stop the collapse and gravity getting ever stronger, a region forms where gravity is so strong not even light can escape – in other words, a black hole. According to Einstein's theory of gravity, if the black hole is sitting there in space, lurking at its centre will be a region of literally infinitely strong gravity, a "singularity". If, however, the black hole is rotating (which, having been formed from a spinning star, it almost certainly will be), then things are rather different. In place of a singularity, it will harbour a "throat" at its centre, into which matter vanishes and from which it is coughed up again somewhere else in the universe. The resulting outpouring of matter would be a white hole: a geyser of matter and energy.

So where are the white holes? On the face of it, they are all over the universe, which teems with evidence of matter and energy pouring into space (indeed, even the Big Bang itself shows signs of having been a white hole). In the mid-1970s, however, an American theorist showed that white holes would be converted into black holes the instant they appeared, as surrounding dust and gas falls on to them and snuffs them out. So, despite being an elegant mathematical possibility, it looks like ugly physics stops us from seeing white holes.

❓ If the sun is made from hydrogen, why doesn't it explode?

The short answer is that the sun is constantly exploding but is prevented from blowing apart by its own colossal gravity field. Inside the sun, hydrogen atoms are crushed together under gravity with such pressure that their nuclei fuse together, triggering the release of energy. The outward radiation pressure and the inward pull of gravity have been in perfect balance for billions of years but after about 8 billion years the fuel will run out.

❓ How fast is the sun burning up its fuel?

The sun is so vast that it can sustain a loss of four million tons a second for billions of years without noticeable effect; certainly there's no chance of measuring it directly. The figure comes from Einstein's famous equation $E = mc^2$ plus measurements of the rate at which the sun emits energy. The sun has been emitting energy for billions of years through nuclear fusion, in which hydrogen nuclei are smashed together so violently they fuse, forming helium nuclei and releasing energy in the process. This reaction is incredibly potent: just one ton of hydrogen fuel releases as much energy as a large power station pumps out in twenty years. It needs to be potent, as the sun emits energy at a rate equivalent to 400 million billion power stations working full pelt. Doing the sums, this means that the sun burns through about 600 million tons of hydrogen fuel each second, which is much higher than the four million tons mentioned above, because most of the hydrogen fuel is converted into helium "ash", which stays on the sun. What is lost forever is the energy given off in the process, in other words, sunlight. Using $E = mc^2$ we find that the sun's colossal energy output produces a loss of four million tons of total mass per second which, happily, the sun can continue to sustain for billions of years yet.

❓ What will happen when the sun runs out of fuel?

This seemed to have been settled years ago but recent research has adding some interesting twists. According to the standard story, as the sun runs out of fuel, it will turn into a red giant star, ballooning outwards to engulf the innermost planets. Calculations suggested that the Earth would be burnt to a cinder around 7.5 billion years from now. It now appears that this scenario is too pessimistic. In 2001, astrophysicists at the University of Sussex announced that the calculations underpinning the standard picture had failed to take into account the fact that the ageing and depleted Sun will weigh rather less than it does now, so its gravity will be somewhat weaker, allowing the orbits of the planets to become somewhat larger than they are today. Taking this into account, the team found that the Earth should always manage to evade the Sun's fiery maw. They also found that the dying sun will make two attempts to destroy our planet. The first is scheduled for around 7.7 billion years from now, when the sun will expand to around 120 times its current size, engulfing Mercury and Venus. About 100 million years later, it will have another attempt but the Earth will by then have moved out of its reach forever.

After that, the Sun is expected to collapse in on itself, the swelling heat from its nuclear reactions unable to resist the inward force of gravity. The result will be the transformation of this red giant star into a harmless white dwarf about 10,000 miles across. The sun will still have much of its original mass and will hang on to most, if not all, of its current retinue of planets – apart from those it fried during its death throes.

? Whatever happened to String Theory as an explanation of the universe?

In the early 1980s, this bizarrely-named idea was touted by many of the world's leading physicists (and, ahem, science correspondents) as the long-sought "Theory of Everything", which would encompass all the sub-atomic particles and forces of nature into one over-arching theory. Their excitement stemmed from the unifying nature of treating sub-atomic particles not as points but as tiny, string-like entities. These strings were said to be trillions of times smaller than an atomic nucleus and to exist only in ten-dimensional space, but they appeared to solve some longstanding problems blocking the way to the Theory of Everything. By the late 1980s the initial euphoria had worn off and physicists were left facing some major problems. First and most obviously, we don't live in a ten-dimensional universe, yet no one could find a neat way of shedding six unwanted dimensions to get down to the four dimensions we inhabit. Second and again pretty obviously, there can be no more than one true Theory of Everything and yet physicists found no fewer than five string theories, with no clear way of choosing between them. Finally, string theories appeared to be mere approximations to something better, raising the question of what that something might be.

It is now clear that the problem with string theory was that it wasn't weird enough. Physicists have since discovered that all five string theories are really just different facets of "M-theory", the M standing variously for "Mother", "Magic" or "Mysterious". Its connection with string theory is a bit clearer if one takes the "M" to stand for "Membrane": ten-dimensional strings can then be regarded as being merely the edges of an eleven-dimensional sheet, which is a key idea in the new theory.

A small army of physicists now works full time on M-theory, which seems to be free of the problems of string theory. For instance, it has revealed that there is just one way of reducing all the extra dimensions down to the four we inhabit. M-theory also appears to come in just one form, as one would expect from the true Theory of Everything. Whether it really is the genuine article is still unclear; all one can say is watch this (eleven-dimensional) space.

? But if there are 11 dimensions, where are they?

The notion that we live in a universe with just three dimensions dates back to the seventeenth century and the philosopher René Descartes, who supposedly had the idea while lying in bed one morning. The story goes that he saw a fly on the ceiling, and realised that just three numbers are needed to describe the location of the fly: the distance from two adjacent walls, plus the height above the ground. A fourth dimension – time – was added in the early twentieth century by Einstein, allowing both the location and timing of events to be fixed exactly. The idea that yet more dimensions might exist was first mooted in 1919 by the mathematician Theodor Kaluza, who made the astonishing discovery that adding a fifth dimension led to a "unified" theory of two of the fundamental cosmic forces: gravity and electromagnetism. The question of where this extra dimension could be was answered in 1926 by the mathematician Oscar Klein, who pointed out that it might simply be curled up far more tightly than the other four dimensions and so can't be observed. This isn't quite so outlandish as it might seem. For example, the Great Wall of China is clearly three dimensional when seen close up but from an aircraft flying high above it only its width and length are obvious: one of its dimensions is no longer visible. From space, the wall loses another dimension and becomes just a thin, one-dimensional line. Thus, as long as they are sufficiently small, it's possible that our universe has many more than just the three spatial dimensions we see. The best candidates for unified theories of cosmic forces currently call for the universe to have no fewer than eleven dimensions but the same idea applies: seven are far smaller than an atom and so cannot be observed directly. Quite why these remained small while just four grew to give us the universe we now see is, as yet, a complete mystery.

Chapter 10

A final miscellany

? **How hazardous are bullets fired into the air?**

Scenes of people firing machine guns with joyful abandon are a familiar sight on TV news programmes. Rather less joyful are the potential consequences of such celebrations. A bullet from a Kalashnikov rifle weighs only about 5 g but leaves the gun travelling at over 1500 mph – twice the speed of sound. This gives the tiny bullet the same amount of energy as a brick dropped from the top of St Paul's Cathedral, so no wonder bullets tend to kill people. If there were no atmosphere, a bullet fired up into the air would come back down with this same amount of energy, and patently lethal consequences. However, air resistance makes a big difference and cuts the final speed of the descending bullet to around ten per cent of the muzzle velocity, about 150 mph and its energy down to the equivalent of a brick dropped on your head from a height of 4 ft or so. Experiments conducted with real falling bullets have confirmed that this is sufficient to cause significant injury and there is anecdotal evidence that they can be lethal. The victims are unlikely to be those doing the firing, however. Travelling thousands of feet into the air, the bullets are usually caught by the wind and land as much as a quarter of a mile away from the gun-toting loons who fire them.

? As we breathe out carbon dioxide, why doesn't blowing on a fire put it out?

It's true that carbon dioxide (CO_2) is used in some fire extinguishers, blasts from which will rob small fires of the free oxygen needed to keep the combustion process going. It's also true that we breathe out this gas at concentrations 100 times higher than those in the air we breathe in, but the concentration of CO_2 in our out-breath is still only around five per cent by weight, so when we blow on a fire, we are still supplying it with substantial concentrations of oxygen, so it responds by glowing brighter. Even if we did breathe out pure CO_2, I doubt we could put out more than a small fire, as we would run out of puff before the fire had cooled sufficiently to prevent it bursting back into flame. For really big fires, CO_2 can actually make things worse, as the heat tears apart the CO_2 molecules, turning them into a rich source of oxygen. This is what happened at the notorious Windscale nuclear reactor fire in October 1957. When scientists tried to douse the melting uranium using liquid CO_2, they discovered they were making the fire worse. Realizing their mistake, they simply poured water straight into the core via fire hoses. Fortunately, this cooled the core sufficiently to put out the fire and to avoid a disaster like that which hit Chernobyl almost thirty years later.

? How do karate chops pack such a punch?

The sight of a karate expert smashing through a heap of roof tiles with a single "chop" is impressive, as is the force involved. Studies using targets wired up to record the effect of the blow have shown that a karate chop can deliver an impact force equivalent to half a ton in weight, all focused on to the edge of the hand – more than enough to break roof tiles, bricks or human bones. The secret to smashing things up lies in delivering as much kinetic energy as possible to the target, deforming the material beyond its limit of flexibility. To do this, the hand has to be moving as fast as possible on contact. Karate experts practise blows which finish an inch or so below the point of contact, with their hand reaching maximum speed, around 20 ft per second, just at the point of contact. Exponents of *tamashiwara*, as this kind of demolition work is called, like to have tiles, blocks or bricks supported at their edges, allowing them to flex and break during the karate chop.

❓ Why are cooling towers the shape they are?

As a regular rail traveller between Oxford and London, I have had all too much time to ponder this question, as the train crawls past the giant cooling towers of Didcot power station. I had thought that the principal explanation for their gently curving shape was structural, giving these huge 300 ft stacks good wind resistance, despite being extremely thin (the walls are typically barely 8 inches thick). This was confirmed by Powergen, who told me that the curving sides help spread the hefty wind-loads around the structure and down into the ground. As a bonus, the narrowing of the towers towards the top slightly compresses the rising steam inside, thus accelerating it – a bit like squeezing the end of a garden hose. The result of this "Venturi effect" is those huge billowing clouds of water vapour rising from the top, which many people mistake for choking white smoke. Dr Chris Burgoyne of Cambridge University tells me that during the 1930s civil engineers tried to boost the Venturi Effect by designing cooling towers that looked like two paper cups joined together at the base. Lacking the nice curvy profile of the classic design, they turned out to be far less resilient against wind loads.

❓ What is "white" gold"?

Often thought to be an alternative name for platinum, white gold is a mixture of pure gold and the silvery metal palladium (though silver itself is sometimes used in cheaper varieties). While the result is generally less expensive than either platinum or pure gold, it is rather dull-looking, so jewellers often coat it with the brilliantly reflective metal rhodium. This can wear off after a few years, revealing the less spectacular white gold underneath. In short, for anyone wanting a ring that will keep its lustre for years, there's no substitute for pure platinum.

❓ How far can a plane glide if its engines stop?

Happily, a lot further than you might think. Aircraft can fly like gliders if all the engines fail; just how far they can glide is dictated by their "Lift-to-Drag" (L/D) ratio, which measures the relative amounts of lift and drag force generated by the wings. The bigger the ratio, the more fuel-efficient the aircraft, the longer its range and the further it will continue to fly if its engines pack up. The precise ratio varies somewhat with speed, dropping dramatically close to the speed of sound but it's typically around 22 for straight-winged gliders, 16 or so for a conventional passenger jet, down to a mere 8 for swept-wing supersonic aircraft like Concorde. In each case, the horizontal distance they can glide can be estimated by multiplying this ratio by the aircraft's altitude when the engines fail. So, for example, if they fail at 32,000 ft – about 6 miles – a passenger jet should be able to reach another destination almost 100 miles away.

So much for the theory – though for once, the reality is not so different. In August 2001, an Canadian Air Transat Airbus 330 flying from Toronto to Lisbon suffered a catastrophic loss of fuel. After noticing the fault, the pilots diverted to an airfield in the Azores but both engines failed while they were still 85 miles away. Fortunately, their aircraft had a L/D ratio of around 16 and the failure took place at 34,500 ft, which meant the aircraft was able to glide to the runway. The pilots made a successful high-speed landing and just a dozen passengers suffered minor injuries.

❓ How big is the "double helix" of DNA?

Astonishingly big: the DNA molecule packed into each

of our cells is around 20 atoms across but about 4 ft long. Tightly coiled up into chromosomes, it can only be seen with an electron microscope but if all the DNA in a single human body were laid end to end, it would reach from the Earth to Pluto ten times over.

? When did Greenwich become the centre of every world map?

Most atlases put Britain at the centre of the world, with the Greenwich Prime Meridian running through the grounds of the Royal Observatory in south London separating west from east. Yet medieval maps such as the thirteenth century *Mappa Mundi* are often centred about the city of Jerusalem because the Book of Ezekiel says the Lord Himself "set her in the centre of nations". The Greenwich Prime Meridian became the official centre of the world only as recently as 1884 when an international conference recognized that the Royal Navy's charts, centred on Greenwich, were so respected and widely-used that it made sense to make all other maps conform to the same prime meridian.

? How did canal tunnel builders achieve such accuracy underground?

The standard method was to lay out the direction by standard surveying methods on the surface and then sink shafts down into the hill to the required depth. Digging would then start from either end and from the shafts, following instructions about which direction to take from the surface. It usually worked pretty well, but was not infallible. The builders of the 1,300 ft long Saltersford Tunnel on the Trent and Mersey Canal near Northwich, Chesire, didn't quite succeed in meeting in the middle, with the result that there is a kink in the tunnel.

❓ What causes the mirror-like mirages on roads in summer?

During the day, the surface of a road mops up the heat of the sun and often ends up much hotter than the air above it. The result is an "inversion layer", in which the air in contact with the surface of the road is less dense than the air above it – a reversal of the usual situation. The optical properties of the air are also affected by the temperature contrast, causing light rays from the sky that would normally run straight into the road to swoop down then bend up again and go straight into our eyes. The inversion layer thus allows us to see the sky without having to look up, a peculiar state of affairs which our brains interpret in much more familiar terms: as the reflection of the sky in water apparently lying on the road.

❓ If an object is dropped in a moving vehicle, why doesn't it move backward?

It seems obvious that an apple dropped on a train should hurtle backwards. After all, in the 0.5 sec it takes to reach the floor, a train doing 125 mph will have moved forward 90 ft. The reason it doesn't is that it is not only the train that is doing 125 mph: so is everything else connected with it, including the apple. So when it's dropped, the apple is still going forward at 125 mph, which ensures it doesn't end up flying backwards and killing passengers who fail to duck. Things become altogether more interesting if the speed of the train changes as the apple falls: then it really will travel backwards, because once it begins to fall, it has no way of increasing its speed to keep up with the accelerating train. If you didn't know the train was accelerating, you might even be led to think that a mysterious force had grabbed hold of the apple and moved it backwards. It was this kind of reasoning that led Einstein to his radical new way of thinking about the force of gravity, which ultimately became the General Theory of Relativity – but then, he was a genius.

? How does "stealth" technology work?

The technique of using special materials to combat the risk of radar detection was pioneered by German submarine designers in the Second World War. During 1944, concern about mounting U-boat losses led to the development of the snorkel, which allowed the submarines to operate on diesel engines while still submerged. German naval commanders believed Allied aircraft fitted with sensitive radar could detect the snorkel and so they developed a special rubber compound designed to mop up radar energy and minimize reflections. It was not a great success, as the sea quickly stripped the coating off. Modern stealth technology first made headlines in 1989, when the US Air Force used F-117A Nighthawk stealth aircraft during the invasion of Panama. Ordinary aircraft bounce back much of the radio wave energy that hits them when they are struck by a radar beam, making them easily detectable to anti-aircraft defences. The F-117A reduces its detectability in two ways. First, its fuselage is coated with layers of carbon or ferrite particles, which mop up a lot of the radar energy that strikes it, turning it into heat. Even the cockpit canopy is coated with transparent, radar-absorbing indium-tin-oxide to prevent the pilot's helmet producing reflected energy. Second, any radar energy that is reflected back is minimized by the angular shape of the F-117A's fuselage, which avoids right angle junctions that can produce strong reflections. The result is a radar image purported to be about the size of a small bird.

? Why is a ship's wash always white, no matter what the colour of the sea?

The turbulence created by the ship's propulsion system and movement through the sea generates a myriad of bubbles. These trap tiny pockets of air in a spherical skin of water molecules, which scatter any light that strikes them. As the light which strikes them is white, we see the result as a roiling mass of whiteness, visible from all directions.

? Why do wagon wheels go backwards in movies?

It's a symptom of the fact that movies are typically shot at the rate of 24 frames per second. If a wagon wheel is turning at the rate of 24 spokes a second, the position of a spoke in one frame will have been taken up by another spoke in the next frame, giving the appearance that nothing has changed and the wheel isn't moving at all. But if the wheel is moving slightly slower than 24 spokes a second, each spoke does not quite have enough time to get into position by the time the next frame is taken and instead appears slightly behind the spoke that appeared in the previous frame. The result is a wagon wheel that appears to be running backwards.

? Why does helium make voices sound squeaky?

Breathing in helium allows anyone to do brilliant Donald Duck impressions for a few seconds. Helium is about seven times less dense than air and, as an inert gas, is made up of individual atoms rather than molecules. This leads to sound travelling through helium around three times faster than through air, which raises the resonant frequencies our throats can create when we speak, resulting in a squeaky voice. A word of warning, though: helium is not air and though it is very light, it could still lead to light-headedness and suffocation if prevented from escaping from the lungs.

? Why does rifling in a gun barrel enable the bullet to travel further?

Invented in the early sixteenth century by European gunsmiths, the spiral grooves inside the barrel cause the bullet to spin as it races out under the explosive charge, reaching rotation speeds in excess of 150,000 revolutions per minute. In just the same way as a gyroscope resists attempts to shift its spin axis, the spinning bullet is much more stable in flight and less prone to tumbling. The result is better accuracy, lower air resistance and improved range. The same effect is put to work on firework rockets, whose plastic fins are twisted slightly to produce spinning during their ascent. Quarterbacks in American football also exploit the spin-stability effect, deliberately spinning the pointed ball as it leaves their hands and travels up-field, reaching its target with often astonishing accuracy.

? Do blind people have dreams?

Yes – although their precise nature depends on when they became blind. People who are blind from birth experience dreams based around bodily sensations, while those who become blind in later life use the store of visual imagery built up during the time they were sighted as the basis of often very vivid dreams. According to Professor J Allan Hobson of Harvard Medical School, author of *Dreaming: an introduction to the science of sleep* (Oxford University Press, 2003), some are even able to use dreams to re-live specific experiences such as family gatherings at will.

? How did people describe years before the invention of BC and AD?

The terms BC and AD, or as is now preferred in modern texts, BCE (Before the Common Era) and CE (Common Era), stem from the work of Dionysius Exiguus (Denis the Lesser), a monk who worked in Rome over 1,400 years ago. In 532 CE Dionysius proposed a Christian calendar with a start date set by an event of great religious significance. All very straightforward but as so often with matters calendrical, we soon get into an unholy tangle; what follows is based chiefly on Dr Duncan Steel's authoritative work *Marking Time* (Wiley, 2000).

First, Dionysius did not do the obvious thing and start the new calendar from the birth of Christ but chose to begin with the Incarnation – that is, the time of the conception of Christ – and used this to set 1 CE as taking place the following year. Historians now believe that this choice was incorrect, as the conception of Christ took place several years before. Whatever: Dionysius ploughed on, referring to all subsequent years as *anno ab Incarnatione*, meaning the year of the Incarnation. The term *Anno Domini* (the year of our Lord) did not appear for another 200 years (it seems to have been first coined by the Venerable Bede) while BC is a more recent invention, first appearing some time during the seventeenth century.

Despite his labours, Dionysius did not live to see his system widely adopted, and for many years after his death dates were still reckoned on the basis of a Roman system, whose start date was set by the establishment of the city of Rome. According to the Roman scholar Marcus Terentius, this took place in 753 BCE, so that subsequent years were prefaced by the term *anno urbis conditae* (AUC – "in the year of the foundation of the city"). You might expect the Romans to have referred to the murder of, say, Julius Caesar in what we call 44 BCE, as an event that took place during AUC 710. Alas, no: for reasons best known to themselves, they eschewed dates in favour of an unwieldy system based on the names of consuls. Thus 44 BCE was known as the first and fifth years of the consulship of Mark Antony and Julius Caesar respectively. I think we should all give thanks that this system of reckoning time went the same way as the Roman Empire.

? How do people sleep so near the edge of the bed without falling out?

All parents know the dangers of very young children falling out of bed and indeed the ability to stay in the bed does appear to be acquired over time. Video footage of sleeping toddlers shows that they are happy to sleep in any orientation, even including face down, so they can easily complete a full roll during the night and end up on the floor. Older children and adults, on the other hand, seem to have learned both to avoid the face down posture and to retain a sense of where they are in the bed, helping them to stay there (in the case of teenagers, all day).

❓ Are conjoined twins necessarily of the same sex?

Formerly known as Siamese twins (after a celebrated pair, Chang and Eng, who were born in Siam, now Thailand, in 1811), conjoined twins are identical twins formed from a single fertilized egg that has undergone incomplete separation. As a result, they have identical DNA and are always the same sex. For unknown reasons, female conjoined twins are over twice as common as male ones.

❓ Why was the UK's 1960s experiment of leaving the clocks on British Summer Time abandoned?

Anyone who has ever forgotten to put the clock back at the end of October has missed out on that rarest of things: a guilt-free extra hour in bed (a handy mnemonic, incidentally, is that the clocks "spring" forward an hour in the Spring, and "fall" back again in the Fall). But in 1968 the British were denied an end-of-October lie-in by order of HM Government, which had set up a three year study of the effects of staying on British Summer Time right through the winter.

I can still recall tramping off to school in what still seemed the dead of night but being glad of the extra hour of sunlight for evening games of footy. However, as the winter drew on, reports emerged of children being killed by juggernauts as they made their way to school through the gloom. Compared to previous years, there seemed to be carnage on the roads and statistics later confirmed that there had indeed been an increase in the number of people killed and injured in the mornings. So even before the official end of the experiment in 1971, politicians had made clear Britain would be reverting to the old system of dark winter evenings.

It is a decision that has since cost thousands of lives. For while there were indeed more deaths and injuries in the mornings, there was a far greater reduction in evening casualties, producing a net benefit of around 2,500 fewer casualties, including several hundred lives. Among the explanations put forward was that the brighter evenings helped motorists get home safely, despite being tired at the end of the day and having to deal with more children on the streets. The problem for the politicians was that the gains were entirely theoretical, for no one could say precisely which children had been spared serious injury or death in the lighter evenings. On the other hand, it was all too easy to identify toddlers who had been run over in the darker mornings.

We might like to think we live in a rational, evidence-based society but in reality, true life tragedies are always more persuasive than theoretical good news.

❓ How can a yacht travel faster than the wind?

There is something undeniably odd about a yacht doing 25 knots while sailing into a 15 knot wind but it helps to think about the relative speed of the boat and the breeze. Clearly, a yacht with the wind directly behind it and no current cannot travel faster than the wind: if it did, its sails would outpace the very thing which is supposedly pushing it forward.

The secret to travelling faster than the wind lies in orienting the yacht so that its sails are set obliquely to an oncoming breeze. By travelling "close to the wind", the yacht's sail deflects the wind as it passes by. By Newton's laws, the resulting change in direction of the moving mass of air generates a force on the yacht, which responds by accelerating. This continues until the force propelling the yacht matches the drag forces slowing it down, at which point it settles down to a "terminal speed", which, for some high-performance catamarans, can be double the wind speed. Even higher speeds can be achieved if the drag forces are reduced, for example by taking the yacht out of the water and putting it on skis. It's said that some ice-yachts skimming across frozen lakes with minimal drag have travelled eight times faster than the prevailing wind.

? **Why do people talk so much? Is there any evolutionary advantage?**

The astonishing popularity of gossip has prompted some academics to wonder if there is more to it than meets the ear. Professor Robin Dunbar, an evolutionary psychologist at the University of Liverpool, has argued that gossiping does for humans what picking nits out of fur does for other primates – namely, it provides a source of social cohesion. In his book *Grooming, Gossip and the Evolution of Language* (Faber, 1996), Dunbar points out that monkeys spend hours grooming each other. Despite its obvious pleasures, when it comes to selecting grooming partners primates are rather picky. They follow strict hierarchies; close friends and relatives get five star treatment, while casual acquaintances get a perfunctory brush-through. Animal behaviourists have found they can work out the social structure of monkey colonies by observing who is grooming whom.

Professor Dunbar makes a plausible case for humans using gossip for the formation and maintenance of social structures. He argues that we evolved language principally to witter on about nothing, allowing us to maintain social cohesion far more effectively than by grooming. Instead of sitting for hours preening each other, humans can gossip while doing other things and, moreover, in far larger groups.

The idea that we should squander such a precious ability on tittle-tattle seems absurd; most mainstream theories assume that language emerged in order to convey complex information about, say, how best to kill a mammoth. Dunbar points out, however, that social cohesion is extremely important and analysis of human conversations shows that about two-thirds of the time is devoted to chatter about social matters. Conclusions based on what we do now hardly make compelling arguments for what we did long ago, however; we modern humans may just have more time to stand around gossiping than our prehistoric forebears. Still, the idea that we gossip to maintain cliques seems perfectly plausible to me. It might also explain why tabloid newspapers devote so much space to trivia.

❓ How do we know that every fingerprint is unique?

This is a rather scary example of how a belief can harden into certainty with the help of a little mathematics and a lot of bluster. The idea that fingerprints are unique goes back centuries and seems to have been based on nothing more than the notion that something so complex cannot be replicated in every detail. The use of fingerprints as a forensic method has its origins in a monograph published in 1892 by the polymath Sir Francis Galton. With characteristic thoroughness, Galton considered all the aspects of fingerprinting needed to turn it into a science, including an estimate of the probability of two people sharing a fingerprint. To do this, he established the size of a patch of fingerprint whose pattern he could identify correctly fifty per cent of the time. Combined with the number of such patches making up a typical fingerprint, Galton estimated that fingerprints were sufficiently different to give odds of around 1 in 64 billion of any two matching by fluke. As this comfortably exceeded the population of the world, Galton concluded that fingerprints are essentially unique. Worryingly, Galton's experiment involved fewer than 100 fingerprints and his probabilistic argument is far from watertight. Yet, as Professor Stephen Stigler points out in his history of statistical concepts *Statistics on the Table* (Harvard University Press, 1999), the prospect of identifying miscreants by fingerprints proved irresistible and by the 1920s standard texts were stating the uniqueness of fingerprints as fact.

For decades Galton's assumption went essentially unchallenged, further enhancing its reputation. The lack of challenges is hardly proof of infallibility – merely of the ability of "scientific" evidence to subdue even the most vociferous defence lawyer. But since the late 1990s there have been several successful challenges to fingerprint evidence, though these have centred chiefly on mis-identification and planted evidence. The consensus remains that fingerprints are essentially unique, despite there being no hope of proving it beyond doubt.

Some suggestive supporting evidence has emerged from studies of the biochemical processes involved in fingerprint formation. Computer models of these processes by Professor James Murray of the University of Washington and his colleagues have recreated the characteristic patterns of ridges. The models also show that even the smallest difference in starting conditions can utterly change the end result. As there is always a random element involved in living processes, this strongly suggests that no two people will have precisely the same fingerprints. So Galton may have been right after all – but more by luck than hard science.

❓ Why do mints make your breath feel cold?

We sense hot and cold through the reaction they trigger in nerve cells, with proteins on their surface altering the flow of ions in and out according to temperature. Researchers at the University of California in San Francisco recently discovered that one such protein is also affected by menthol, the chemical in mints. Sucking a mint releases enough menthol on to nerve cells to trigger the same effect as a cold drink and so creates the illusion of breathing in cool air.

❓ What exactly is suction?

Physics teachers often declare that suction is a myth. This is hard to credit when you use a straw to drain your fizzy drink, yet they're right: suction does not really exist. We are not so much sucking the drink up the straw as creating conditions where the surrounding atmosphere can push the liquid up it instead. This requires that we reduce the pressure at the point towards which we want the liquid to go, which we do by extracting air from the top of the straw. While this involves sucking out the air, the upward movement of the liquid is actually the result of having higher air pressure at the bottom of the straw than at the top.

For something that doesn't really exist, "suction" is an impressively powerful effect, underpinned by the fact that, at sea level, atmospheric pressure amounts to a force of around 15 lbs of weight acting on every square inch. Creating even a partial vacuum thus gives one access to a surprisingly potent force, as Otto von Guericke demonstrated in the mid-1650s. Using a simple air pump, he created a vacuum between two metal hemispheres 1 ft or so in diameter, held together by nothing more than a bit of grease. He then hitched teams of horses to each hemisphere and set them pulling in opposite directions. They completely failed to separate the hemispheres. While there was nothing in the hemispheres but a vacuum, the pressure of the whole Earth's atmosphere was pushing on their outer surfaces, clamping them together with the force of several tonnes of weight.

? How does a siphon persuade water to propel itself upwards?

I've often had cause to exploit the siphon effect – chiefly in cafés, as a means of amusing my offspring while waiting for their meals to arrive. Using a bendy drinking straw between two glasses, one held higher than the other, a quick suck on the lower end of the straw has allowed me to transfer fizzy liquids up and over the edge of the higher glass and into the lower, apparently against gravity. What makes the effect so intriguing is that the the liquid continued to flow without any mechanical aid, even after one has stopped sucking. One way to understand it is to think of the column of drink in the straw as a fine chain draped over a peg: if one end is lower than the other, gravity can drag the chain over the peg. In a siphon, the initial suck on the lower end drags the drink up and over the top of the bend (though, as the sucking creates a vacuum, the liquid is actually being pushed up the straw by the atmospheric pressure at the other end). Gravity then does the rest. Liquid isn't really like a chain, so why doesn't the column break apart as it flows up the straw? The answer is air pressure: if a break did develop in the liquid column, the space in between would contain a vacuum that would quickly be re-filled by liquid pushed into it by atmospheric pressure. Amazingly enough, the resulting column can stay intact and flow upwards to heights of over 30 before descending again.

? Is it true that, as the song says, "There are more questions than answers"?

A Top Ten hit for Johnny Nash back in 1972, I believe – but I digress. There are clearly many more questions than we presently have answers for: is there a God? How did the universe begin? Will Man United win the Premiership? This is not to say such questions do not have answers: Christians, cosmologists and United supporters clearly believe they do.

The more interesting issue is whether there are questions for which answers simply do not exist. In the 1930s, the mathematicians Alonzo Church and Alan Turing looked into this question and independently reached the same conclusion: Johnny Nash is right: there really are more questions than answers.

Many readers will, I think, share my perplexity at how one could think of a way of proving so general a statement. Apparently, the 23-year old Turing hit upon his strategy while lying in a meadow at Grantchester, near Cambridge, in the early summer of 1935. His method not only solves the problem but also casts light on what it takes to be considered a mathematical genius.

Older readers will recall that arithmetic was once divided into two types: mental and mechanical; the latter being tedious sausage-machine stuff like long division. Turing's stroke of genius was to take the term "mechanical" literally. He showed that questions could be cast into a form allowing them to be dealt with by a machine, which cranked through all the stages automatically, spat out the answer and stopped. Not such a radical idea in these days of the computer but this was years before the first computer had (with help from Turing) been built. Turing cast the conundrum of whether all questions have answers in concrete terms: was it possible to say whether such a machine would come to a stop for any given input? He imagined using the machine itself to try to answer this "Halting Problem" and proved that it became trapped in a circular contradiction.

By showing that the question "Does the machine halt for any given input?" led to endless circular contradiction, Turing had shown there is at least one question with no answer. Since then, mathematicians have identified many other such questions, all of which can be boiled down to Turing's Halting Problem and thus cannot be answered. As a simple example, consider the following: "This statement is false". So – is it?

? Why do most wind turbines have three blades rather than four?

Or indeed lots of blades, like those farm water-pump turbines familiar from Westerns. One reason for preferring three blades over four is cost: for any given size of turbine, it's cheaper to have fewer blades. Aesthetics is also an issue; people find three-bladed slightly less offensive than four-bladed leviathans.

There is a more serious problem for any turbine fitted with an even number of blades. When one blade is pointing vertically upwards and feeling the full force of the wind behind it, its partner directly below is shielded by the vertical tower – creating huge imbalances and severe stability problems.

❓ Is there a universal definition of "left" and "right"?

No less a philosopher than Immanuel Kant sensed that there were hidden depths in this question, though I suspect even he would have been surprised by the ultimate answer. In a short essay published in 1768, *Concerning the Ultimate Ground of the Differentiation of Directions in Space*, Kant pointed out that there was something odd about left- and right-handedness. Usually, when we say two objects are "different", we mean that measurements of their properties will produce different values, for example greater width or lesser weight. Showing that they are different is thus usually pretty simple. Yet how does one describe a difference in handedness? We can see that our left and right hands are different but how would you explain that difference to a race of aliens unable to see them? Left- and right-handedness are clearly fundamental properties of objects (indeed, life itself depends on the handedness of DNA and amino acids) so surely there must be a universal standard against which we can define left and right?

There is, and it lies buried in the design of the cosmos. Glimmerings of its existence first emerged in 1918 when the mathematician Emmy Noether demonstrated a deep connection between the mathematical notion of symmetry and conservation laws, such as conservation of energy, in physics. Noether's Theorem forms a bridge between mathematical properties that don't change – "symmetries" which keep things identical after, say, being seen in a mirror – and unchanging physical properties. It also gives a clue about where to find a universal standard for handedness. For if there is no absolute way of telling left and right apart, this means our universe possesses a fundamental left–right symmetry and so by Noether's Theorem, a specific conservation law of physics must always hold true. In the mid-1950s, physicists found this wasn't so: particles emerging from radioactive atoms violate a conservation law related to left–right symmetry. Thus we come to the rather astonishing conclusion that there is indeed a universal standard of left and right but it can only be found by studying the paths of certain sub-atomic particles. That Kant was able to use reasoning alone to understand that the answer to so simple a question might prove so fundamental is surely testimony to his genius as a philosopher.

? Why is carbon monoxide (CO) so deadly, when half of it is oxygen?

Questions like this made me give up chemistry at the first opportunity, only to regret having done so years later. They seemed proof that chemistry didn't make sense. Yet there is surely something fascinating in a subject that can explain such paradoxes as why oxygen-rich carbon monoxide is so lethal – or why common salt isn't, given that half of it is made from chlorine which killed thousands of troops in the First World War.

The resolution of such paradoxes lies in the chemical bond created by the clouds of electrons surrounding each atom. In the case of salt, the strong bonds between its sodium and chlorine atoms ensure the latter do not form molecules that then combine with moisture to produce the hydrochloric acid and oxygen radicals responsible for the lethal effects of chlorine. The danger of carbon monoxide stems from its inter-action with haemoglobin, the protein responsible for transporting oxygen around the body. Haemoglobin has sites on its surface for carrying oxygen molecules; unfortunately, these sites are even better suited to any molecules of carbon monoxide that happen to be about. The result is a frighteningly effective form of molecular-level suffocation: CO concentrations of just 1 part per 1000 of air can prove lethal. Incidentally, carbon dioxide can also be lethal, as was shown in 1986 when 1,700 people died in Cameroon as a result of the release of CO_2 from Lake Nyos. As CO_2 does not share the affinity for haemoglobin of its monoxide relative, fatal concentrations are around 100 times higher.

? How do those "dipping bird" toys keep going without a source of energy?

The secret to the bird's seemingly perpetual motion lies in the liquid sealed in its long, thin neck: methylene chloride. This very volatile compound turns from liquid to vapour with very little heat; indeed, ordinary room temperatures suffice. The resulting vapour rises into the bird's head, which is covered by a felt-like material. Keeping this damp promotes evaporation, which cools the head and allows the vapour to turn back into liquid. That, in turn, lowers the pressure in the bird's nether regions, driving fluid up into the bird's head, until it becomes top heavy. Eventually, the bird leans right over and dips its beak in the water. The liquid in its head then flows back into the base and the whole process starts again. So it's not real perpetual motion: the bird's energy source is the warmth of the room and it will keep doing its little trick as long as it can moisten its head to refresh the evaporation effect.

? How does the "memory metal" used in spectacle frames remember its shape?

The bizarre ability of certain metals to untwist themselves and return to their original shape was first seen in an alloy of gold and cadmium by the metallurgist Arne Olander in 1932. It took another thirty years for researchers to find the first cheap and convenient example of a "shape memory alloy" (SMA): Nitinol, a combination of nickel and titanium, found by a team at the US Naval Ordnance Laboratory (and named after the chemical abbreviations of the two metals, plus the initials of the laboratory). The secret of SMA lies in the arrangement of the metal atoms. Like water turning to ice, metals undergo fundamental changes in the arrangements of their atoms as they cool from the molten state. As in many alloys, the atoms in SMAs re-arrange themselves during cooling from a relatively strong, cubic pattern into a weaker, skewed pattern. Unlike ordinary alloys, however, the atomic bonds in SMAs allow them to return to their original, stronger, cubic pattern if reheated. So components made from SMAs start off in the strong form, retaining their shape as they cool into the weaker form. If they subsequently become bent, a little heating returns them to the stronger form in which they were created, altering their atomic bonds and restoring the original shape.

It doesn't take much heating to "remind" SMAs of their original shape (a hair-dryer will do). SMAs used in spectacle frames will restore themselves without any help: when exposed to a load, such as someone sitting on them, the alloy takes up the weaker, skewed atomic arrangement; when the load is relieved, the atoms spring back into their original, stronger, arrangement, restoring the shape of the frames. While undoubtedly a marvel of materials science, anyone keen to show off their fancy SMA frames to their friends should beware: they have a far lower fatigue strength than conventional metals. In other words, if you ask SMAs to perform their trick too often, you'll break them.

? Why can we not see clearly underwater without the aid of a mask or goggles?

The blurring effect is caused by our eyes being unable to focus the incoming light correctly on to the retina. Whenever light passes from one medium to another, its speed and direction change slightly. Our eyes are tuned to cope with light that enters them after passing through air, which ensures that the resulting bending creates a sharply-focused image. If there is water touching our eyes, the level of bending is altered and our eyes can no longer bend the light by the right amount to achieve a sharp image.

❓ Can light be created without heat?

We normally associate light with heat: fires, candles, the sun. Standard incandescent light bulbs should more properly be called heat bulbs, as ninety per cent of the electrical energy fed into them emerges as heat. There are ways of generating light without heat however, and they work by exploiting means other than heat to persuade electrons to jump into higher energy states within atoms, from which they drop back down, giving off their energy as light. Collectively, such processes are known as luminescence, and within that broad definition are various means of generating "cold light". For example, fluorescent compounds will give a brief burst of light if exposed to the right stimulus but need constant stimulation to keep going. Old luminous watch dials are a case in point: they used compounds like zinc sulphide, whose relatively brief fluorescence was stoked using the radioactive decay of radium (which led to deaths among those who made the dials). Today's luminous watches use compounds based on "rare earths" like europium, whose electrons respond to ordinary daylight, pumping out astonishingly bright cold light all night long.

As so often, nature got there first: many living organisms such as fish and fireflies have evolved chemicals which emit cold light with impressive efficiency. Even some fungi manage it, giving off an eerie glow as they decay in forests at the dead of night.

❓ Why does spitting into a diving mask stop it from misting up?

It's the same effect as dabbing your finger with a bit of soap and wiping it across a steamed-up mirror. Saliva helps breaks the surface tension of the condensation forming on the inside of the mask, so that it spreads out into a relatively distortion-free flat layer you can see through, instead of a myriad of tiny, round droplets.

? What causes the tiny streams of bubbles rising in a glass of champagne?

This is one of those "trivial" questions that led to a Nobel Prize. The bubbles in champagne and fizzy drinks are in general filled with carbon dioxide, which emerges from the liquid after having been crammed into it under pressure. As the molecules of CO_2 make their escape, some hit the side of the glass and get trapped inside tiny cracks or under specks of dust. If enough gas accumulates at one of these "nucleation points", it forms a bubble, which grows until it breaks away and floats up to the surface, leaving the nucleation point free for another bubble to form. The result: a constant stream of bubbles apparently rising from nowhere. In 1952, the physicist Donald Glaser was pondering bubbles rising in his glass of beer when he realised that the same process could reveal the presence of invisible sub-atomic particles, which would leave a trail of bubbles as they zoomed through liquid hydrogen. This led him to invent the "bubble chamber" and won him the 1960 Nobel Prize for physics.

? What is the strongest acid?

The most feared of the acids regularly used in laboratories is hydrofluoric acid (HF), used for cleaning metals. A fuming, colourless liquid, HF is appallingly corrosive, as many horror stories attest. In one incident, a technician in a laboratory spilled a small cupful of HF on to his thigh; despite immediately dowsing himself in water and being rushed to hospital, he still lost his leg. Even then, the reaction between the HF and the calcium in his bones did not stop and he died fifteen days later. Oddly enough, despite its ferocious action on metals and living tissue, HF can be stored in bottles made of certain types of plastic.

There are even more corrosive specialist acids. By mixing together various horrendous substances, chemists have come up with a mixture of antimony pentafluoride, fluorosulphonic acid and sulphur trioxide, whose acidity is far greater than any other known compound; so great in fact, that no one knows for sure how acidic it really is.

❓ Why are some solids like glass and perspex transparent?

Anything that is transparent allows light rays to get through the thicket of atoms packed within it more or less unscathed. As a single grain of salt contains around twenty billion billion atoms, it's no surprise that light usually fails to find a route through most solids and ends up being scattered, reflected or mopped up by the clouds of electrons surrounding each atom. The reason light can make it through substances like glass is that their molecules only mop up light of wavelengths shorter than those of visible light (including, thankfully, cancer-causing ultraviolet light). They are also smooth and amorphous and so have an internal structure lacking anything comparable in size to the longer wavelengths of visible light. These rays can thus get through with minimal scattering and loss of energy, allowing us to see objects through them clearly and brightly. For many solids, there is no way to alter their molecular or internal structure to make them transparent. One alternative is to switch to "light" of a shorter and more penetrating wavelength, such as X-rays.

❓ Why do metal objects feel colder than wooden ones, when both are at the same temperature?

When we touch an object, it's not its actual temperature that we feel but the rate and direction of heat flow between our fingers and the object. For example, in a room at 21 °C, any objects are cooler than our fingers (which are closer to body temperature, around 37 °C). Despite all the objects being at the same temperature, any wooden ones still manage to feel relatively warm, because wood conducts heat hundreds of times less well than metal and so the loss of heat from our fingers to the objects is much slower. With objects that are hotter than us, the effect works the other way around, making wooden objects feel relatively cool. So, for example, a wooden bench that's been in the sun all day won't feel as hot as a car bonnet, even if both are at the same temperature, because the rate at which we gain heat from the wooden bench is much lower than it is from the metal of the car.

Index

Index

Index

Index

Index

suction, 217–18
sun, 149–51, 166, 174–5, 177, 198–200
sunburn, 51–2
Super Glue (mythyl cyanoacrylate),
    6–7
superconductivity, 120
surveying methods, canal tunnels,
    207
Swartz, Professor Clifford, 17
syn-propanethial-S-oxide, 28
Szekely, Dr Tamas, 133

Taylor-Jones, Sonja, 124
tea-making, 3–4
Teflon (polytetraflouroethylene), 21
television, 9–10
temperature
    air turbulence and heat, 104
    Celsius and centigrade, 111
    cold viruses and, 32
    Earth's interior, 157
    freezing point, 4–6
    heating water, 4
    hottest part of day, 102
    light bulbs and, 18
    light without heat, 225
    measurement of speed of Earth,
        187
    relative heat flow, 227–8
    snowfall and, 105–6
    sound and, 111–12
    ultimate maximum, 192
    wind speed, 107
Tennant, Colin, 132
Thomas, Professor Lyn, 20
Thomson, Joseph, 1
Thorp, Dr Edward, 97
ticklishness, 65, 129
time
    before invention of BC and AD,
        211–12
    Daylight Saving/Summer Time,
        91, 213–14
    Easter dates, 89–80
    Greenwich Mean Time, 151
    outward and inward journeys, 3
    quartz crystal clocks, 16
    rainfall volume and, 101
    spent searching, 20–1

of sunrise and sunset, 174–5
    of tax year, 8
toothpaste, 14
transmission of disease, 39, 49, 50–1
transparent solids, 227
transpiration, 128
Travisano, Professor Michael, 60
trees, 128, 134–5
Trefethen, Lloyd, 56
Tremaine, Professor Scott, 178
Tryon, Edward, 188
tsunami, 155–6
Turing's Halting Problem, 219–20

ultraviolet light, 51–2, 139, 158
universe
    Big Bang, 187, 188–9, 190–1, 193–4,
        198
    expansion of, 191–4
    galaxies, 173, 187, 190, 191
    pull of gravity, 185–6
    radius of, 189–90, 195
    String Theory, 200–2
    vacuum state, 188–9
    where it begins, 163
    see also, space

vacuum state, 188–9
vacuum under pressure, 218, 219
Van Allen Belts, 155
velocity, 7–8, 164–5
Venturi Effect, 205
vibration, 137
viruses, 32–3, 50–1
visible universe, 190
vision, 50, 173, 224
    at speed of light, 187–8
    see also, optical effects
visualizing a million, 84
vitamins, 50, 139
volcanoes, 155–7
von Guericke, Otto, 218
von Neumann, John, 122–3, 189

Washoe (the chimpanzee), 135
water, 159, 209–10, 224
    atmospheric, 147–8
    draining away (Coriolis Force),
        56–7

About the Author

Robert Matthews is Visiting Reader in Science at Aston University, Birmingham. He has published pioneering research in fields ranging from code-breaking to predicting coincidences, and won an Ig Nobel Prize for Physics for his studies of Murphy's Law, including the reason why toast so often lands butter-side down. An award-winning science writer, he has contributed to many national newspapers and magazines world-wide, and is currently science consultant for *BBC Focus*. His book *25 Big Ideas: The Science that's Changing our World* is also published by Oneworld.